Nanotechnologies in Green Chemistry and Environmental Sustainability

Nanotechnologies represent a fast-growing market and this unique volume highlights the current studies in applied sciences on sustainability of green science and technology. The chapters include modelling, machine learning, nanotechnology, nanofluids, nanosystems, smart materials and applications, and solar and fuel cells technology. The authors cover simulation, additive manufacturing, machine learning and the autonomous system. Various aspects of green science as well as transdisciplinary topics between fundamental science and engineering are presented. The book is suitable for all postgraduates and researchers working in this rapid growing research area.

Sustainability: Contributions through Science and Technology
Series Editor: Michael C. Cann, Ph.D.
Professor of Chemistry and Co-Director of Environmental Science
University of Scranton, Pennsylvania

Preface to the Series

Sustainability is rapidly moving from the wings to center stage. Overconsumption of non-renewable and renewable resources, as well as the concomitant production of waste has brought the world to a crossroads. Green chemistry, along with other green sciences technologies, must play a leading role in bringing about a sustainable society. The Sustainability: Contributions through Science and Technology series focuses on the role science can play in developing technologies that lessen our environmental impact. This highly interdisciplinary series discusses significant and timely topics ranging from energy research to the implementation of sustainable technologies. Our intention is for scientists from a variety of disciplines to provide contributions that recognize how the development of green technologies affects the triple bottom line (society, economic, and environment). The series will be of interest to academics, researchers, professionals, business leaders, policy makers, and students, as well as individuals who want to know the basics of the science and technology of sustainability.

Michael C. Cann

Nanotechnologies in Green Chemistry and Environmental Sustainability

Edited By
Samsul Ariffin Abdul Karim
Faculty of Computing and Informatics
Universiti Malaysia Sabah, Malaysia

CRC Press
Taylor & Francis Group
Boca Raton London New York

CRC Press is an imprint of the
Taylor & Francis Group, an **informa** business

First edition published 2023
by CRC Press
6000 Broken Sound Parkway NW, Suite 300, Boca Raton, FL 33487-2742

and by CRC Press
4 Park Square, Milton Park, Abingdon, Oxon, OX14 4RN

ISBN: 978-1-032-34132-3 (hbk)
ISBN: 978-1-032-34150-7 (pbk)
ISBN: 978-1-003-32074-6 (ebk)

DOI: 10.1201/9781003320746

Typeset in Times
by Newgen Publishing UK

Contents

Preface

Green chemistry has been developed as a natural evolution to prevent the pollution. There are 12 principles in green chemistry. There is the need for a book that could provide the state of the art of green chemistry especially among the Association of Southeast Asian Nations (ASEAN) countries. This book is a collection of the works that have been conducted by researchers at various universities in ASEAN countries. This book highlights the current state of the studies in applied sciences towards the sustainability of the green science and technology, including modelling, nanotechnology, nanofluids, nanosystems, smart materials and applications, and solar and fuel cells technology. Since the introduction of Industrial Revolution 4.0 (IR 4.0), most countries all over the world are encouraging the industrial partners to utilizing IR 4.0 in all aspects of the product life cycle. This will enable all the processes to be automated. To achieve this, it is important that we have the knowledge and expertise as well as abilities to fully utilize IR4.0. Basically, there are seven pillars under IR4.0. This book consists of 14 chapters that cover simulation and additive manufacturing that result in the *enhancement of the reuse of products or materials; enhancement of the use of renewable materials; improvement of the longevity of a material or device as well as prevention of waste.* Each contributed chapter can be regarded as a self-standing contribution. The main outcome of this book is in line with the United Nations Sustainable Development Goals (UN SDG) that lead to the green chemistry, i.e., SDG No. 7: Affordable and Clean Energy and SDG No. 9: Innovation, Industry, and Infrastructure. It would be a dream that our world would be able to meet and achieve all 17 UN SDGs through the green chemistry.

I would like to thank all the contributors who provided an excellent contribution for this book. I am forever grateful for their commitment and contribution to this book. A special thank you to the publishing staff at Taylor & Francis Group/CRC Press for their kind help. I would also like to acknowledge the Faculty of Computing and Informatics, Universiti Malaysia Sabah, for the financial and computing facilities supports that have made the completion of this book possible.

This book is suitable for all postgraduates and researchers working in this rapid growing research areas.

Samsul Ariffin Abdul Karim
Kota Kinabalu, Malaysia
May 2022

Editor Biography

Samsul Ariffin Abdul Karim is an Associate Professor with Software Engineering Programme, Faculty of Computing and Informatics, Universiti Malaysia Sabah (UMS), Malaysia. He obtained his PhD in Mathematics from Universiti Sains Malaysia (USM). He is a professional technologist registered with Malaysia Board of Technologists (MBOT), No. Perakuan PT21030227.

His research interests include numerical analysis, machine learning, approximation theory, optimization, science, and engineering education as well as wavelets. He has published more than 140 papers in journals and conferences, including three edited conferences volume and 60 book chapters. He was the recipient of Effective Education Delivery Award and Publication Award (Journal & Conference Paper), UTP Quality Day 2010, 2011 and 2012, respectively.

He was a certified WOLFRAM Technology Associate, Mathematica Student Level. He has also published ten books with Springer Publishing, including five books with Studies in Systems, Decision and Control (SSDC) series, one book with Taylor and Francis/CRC Press, one book with IntechOpen and one book with UTP Press.

Recently he has received the Book Publication Award in UTP Quality Day 2020 for his book *Water Quality Index Prediction Using Multiple Linear Fuzzy Regression: Case Study in Perak River, Malaysia*, which was published by SpringerBriefs in Water Science and Technology in 2020.

Contributors

Mochamad Achyarsyah
Foundry Department, Bandung
  Polytechnic of Manufacturing,
  No. 21 Kanayakan Street, Bandung,
  West Java, Indonesia

Aulia Amar
Department of Mechanical Engineering,
  Faculty of Engineering, Universitas
  Negeri Malang, Jl. Semarang No. 5,
  Malang 65145, Indonesia

Aminnudin
Department of Mechanical Engineering,
  Faculty of Engineering, Universitas
  Negeri Malang, Jl. Semarang No. 5,
  Malang 65145, Indonesia

Desi Puspita Anggraeni
Department of Mechanical Engineering,
  Faculty of Engineering, Universitas
  Negeri Malang, Jl. Semarang No. 5,
  Malang 65145, Indonesia

Ananta Ardyansyah
Faculty of Science, Universitas Negeri
  Malang, Jl. Semarang No. 5, Malang
  65145, Indonesia

Nadiya Ayu Astarini
Physics Department, Universitas Negeri
  Malang, Jl. Semarang No. 5, Malang
  65145, Indonesia

Muchammad Riza Fauzy
Department of Industrial Engineering,
  Faculty of Engineering, University
  of Merdeka Malang, Jalan Raya
  Terusan Dieng 62-64, Malang 65156,
  Indonesia

Christian Hadhinata
Civil Engineering Department,
  Universitas Negeri Malang, Jl.
  Semarang No. 5, Malang 65145,
  Indonesia

Samsul Ariffin Abdul Karim
Software Engineering Programme
Faculty of Computing and Informatics
Universiti Malaysia Sabah
Jalan UMS, 88400 Kota Kinabalu
Sabah, Malaysia
Data Technologies and Applications
  (DaTA) Research Group, Universiti
  Malaysia Sabah, Jalan UMS, 88400
  Kota Kinabalu, Sabah, Malaysia

Syafira Bilqis Khoyroh
Department of Mechanical
  Engineering, Faculty of Engineering,
  Universitas Negeri Malang, Jl.
  Semarang No. 5, Malang 65145,
  Indonesia

Djoko Kustono
Department of Mechanical
  Engineering, Faculty of Engineering,
  Universitas Negeri Malang, Jl.
  Semarang No. 5, Malang 65145,
  Indonesia

Afzan Mahmad
Fundamental and Applied Sciences
  Department, Universiti Teknologi
  PETRONAS, Bandar Seri Iskandar,
  32610 Seri Iskandar, Perak Darul
  Ridzuan, Malaysia

Mohd Faridzuan Majid
Fundamental and Applied
  Sciences Department, Universiti
  Teknologi PETRONAS, 32610
  Seri Iskandar, Perak Darul Ridzuan,
  Malaysia

Danialnaeem Emirzan Bin Mardani
Chemical Engineering Department,
  Universiti Teknologi PETRONAS,
  32610 Seri Iskandar, Perak Darul
  Ridzuan, Malaysia

Fajar Muktodi
Department of Mechanical Engineering,
  Faculty of Engineering, Universitas
  Negeri Malang, Jl. Semarang No. 5,
  Malang 65145, Indonesia

Yuke Nofantyu
Department of Mechanical Engineering,
  Faculty of Engineering, Universitas
  Negeri Malang, Jl. Semarang No. 5,
  Malang 65145, Indonesia

Teh Ubaidah Noh
Institute of Bioproduct Development,
  Universiti Teknologi Malaysia

Riana Nurmalasari
Department of Mechanical Engineering,
  Faculty of Engineering, Universitas
  Negeri Malang, Jl. Semarang No. 5,
  Malang 65145, Indonesia

Avita Ayu Permanasari
Department of Mechanical Engineering,
  Faculty of Engineering, State
  University of Malang, Jl. Semarang 5,
  Malang, 65145, Indonesia
Center of Advance Materials and
  Renewable Energy, State University
  of Malang, Jl. Semarang 5, Malang,
  65145, Indonesia

Yanuar Rohmat Aji Pradana
Department of Mechanical Engineering,
  Faculty of Engineering, Universitas
  Negeri Malang, Jl. Semarang No. 5,
  Malang 65145, Indonesia

Diki Dwi Pramono
Department of Mechanical Engineering,
  Faculty of Engineering, State
  University of Malang, Jl. Semarang 5,
  Malang, 65145, Indonesia

Muhammad Mirza Abdillah Pratama
Civil Engineering Department,
  Universitas Negeri Malang, Jl.
  Semarang No. 5, Malang 65145,
  Indonesia

Hakas Prayuda
Department of Civil Engineering,
  Faculty of Engineering, Universitas
  Muhammadiyah Yogyakarta,
  Indonesia

Herlin Pujiarti
Physics Department, Universitas Negeri
  Malang, Jl. Semarang No. 5, Malang
  65145, Indonesia
Centre of Advanced Materials for
  Renewable Energy (CAMRY),
  State University of Malang, Jl.
  Semarang No 5, Malang 65145,
  Indonesia

Poppy Puspitasari
Department of Mechanical Engineering,
  Faculty of Engineering, State
  University of Malang, Jl. Semarang 5,
  Malang, 65145, Indonesia
Centre of Advanced Materials and
  Renewable Energy, State University
  of Malang, Jl. Semarang 5, Malang,
  65145, Indonesia

Andika Bagus Nur Rahma Putra
Department of Mechanical Engineering,
  Faculty of Engineering, Universitas
  Negeri Malang, Jl. Semarang No. 5,
  Malang 65145, Indonesia

Viska Rinata
Faculty of Science, Universitas Negeri
  Malang, Jl. Semarang No. 5, Malang
  65145, Indonesia

Pungky Eka Setyawan
Department of Mechanical Engineering,
  Faculty of Engineering, University of
  Merdeka Malang, Jl. Raya Terusan
  Dieng No. 62-64, Malang 65156,
  Indonesia

Maizatul Shima Shaharun
Fundamental and Applied Sciences
  Department, Universiti Teknologi
  PETRONAS, Bandar Seri Iskandar,
  32610 Seri Iskandar, Perak Darul
  Ridzuan, Malaysia

Nabella Sholeha
Physics Department, Universitas Negeri
  Malang, Jl. Semarang No. 5, Malang
  65145, Indonesia

Sukarni Sukarni
Department of Mechanical Engineering,
  Faculty of Engineering, Universitas
  Negeri Malang, Jl. Semarang No. 5,
  Malang 65145, Indonesia

Agus Suprapto
Department of Mechanical Engineering,
  Faculty of Engineering, University of
  Merdeka Malang, Jl. Raya Terusan
  Dieng No. 62-64, Malang 65156,
  Indonesia

Suprayitno
Department of Mechanical Engineering,
  Faculty of Engineering, State
  University of Malang, Jl. Semarang 5,
  Malang, 65145, Indonesia

Heru Suryanto
Department of Mechanical Engineering,
  Faculty of Engineering, Universitas
  Negeri Malang, Jl. Semarang No. 5,
  Malang 65145, Indonesia

Dewi 'Izzatus Tsamroh
Department of Mechanical Engineering,
  Faculty of Engineering, University of
  Merdeka Malang, Jl. Raya Terusan
  Dieng No. 62-64, Malang 65156,
  Indonesia

Hayyiratul Fatimah Mohd Zaid
Centre of Innovative Nanostructures and
  Nanodevices (COINN), Universiti
  Teknologi PETRONAS, 32610 Seri
  Iskandar, Perak Darul Ridzuan,
  Malaysia
Chemical Engineering Department,
  Universiti Teknologi PETRONAS,
  32610 Seri Iskandar, Perak Darul
  Ridzuan, Malaysia

Siti Nur Azella Zaine
Chemical Engineering Department,
  Universiti Teknologi PETRONAS,
  Bandar Seri Iskandar, 32610 Seri
  Iskandar, Perak Darul Ridzuan,
  Malaysia

1 Introduction

Samsul Ariffin Abdul Karim*
Software Engineering Programme
Faculty of Computing and Informatics
Universiti Malaysia Sabah
Jalan UMS, 88400 Kota Kinabalu
Sabah, Malaysia; Data Technologies and Applications (DaTA)
Research Group, Universiti Malaysia Sabah, Jalan UMS, 88400
Kota Kinabalu, Sabah, Malaysia

CONTENTS

1.1 INTRODUCTION

Greening chemistry has been developed by the experts as a natural evolution to prevent the pollution that maybe coming from human activities or industries. Besides, the United Nation Sustainable Development Goals (UN SDG) [1] also lead to the green chemistry such as SDG No. 7: Affordable and Clean Energy and SDG No. 9: Innovation, Industry, and Infrastructure. The main finding in this book has the motivation to improve our nature that will contribute to the sustainability of the green chemistry. The main challenges are how to implement the Industrial Revolution 4.0 into the green chemistry. To answer this question, there are 14 chapters in this book that provide significant ideas and suggestion to cater the problem.

1.2 SUMMARIES

The following paragraphs provide the main summaries for each contributing chapter. Each contributed chapter can be regarded as a self-standing contribution.

Chapter 2 entitled "Phase Identification, Morphology, and Compressibility of Scallop Shell Powder (Amusium Pleuronectes) for Bone Implant Materials" discusses the effect of sintering time on phase, morphology, distribution of functional groups, and compressibility of calcium carbonate ($CaCO_3$) from scallop shell powder. The sintering condition will affect the morphology (characterized by scanning electron microscope), phase identification (characterized by X-ray diffraction), molecular bonding (characterized by Fourier transform infrared), and also the compressibility of $CaCO_3$. Furthermore, $CaCO_3$ produced from scallop shell can be used as bone implant materials in the form of hidroxyapatite.

DOI: 10.1201/9781003320746-1

Chapter 3 entitled "Simulation for Oil Pan Production against Its Porosity, Shrinkage, and Niyama Criterion" studies the possibility of casting defects that occur in high-pressure die casting (HPDC) products using finite element method (FEM). The use of FEM in metal casting simulations is carried out to predict any porosity and shrinkage defects that may occur due to nonuniform metal flow and the presence of eddies during liquid filling in the mold cavity, and to predict casting defects that will involve the Niyama criterion to see the contours of the formation of casting defects. Additionally, the inexpensive technology with all the parameters in casting simulations will enhance the sustainability of casting products yields.

Chapter 4 entitled "Analysis of the Thermophysical Properties of SAE 5W-30 Lubricants with the Addition of Al_2O_3, TiO_2, and Hybrid Al_2O_3-TiO_2 Nanomaterials on the Performance of Motorcycles" develops nanolubricant to enhance the performance of motorcycle. The addition of nanomaterials to the base fluid changes its thermophysical features such as thermal conductivity, density, and viscosity. Dimensions and shape of nanomaterials, concentration of nanomaterials, viscosity of base fluids, and surfactants used for formulation are the main parameters that affect thermophysical behavior. The research was done by adding nanomaterial Al_2O_3, TiO_2, and the mixture of both Al_2O_3 and TiO_2 in 5W-S30 lubricant. The dyno test is carried out to determine the engine performance of motorized vehicles. Moreover, the efficiency and the effectiveness of motor vehicle engine performance will increase by the influence of nanolubricant.

Chapter 5 entitled "Heat Transfer Rate and Pressure Drop Characteristics on Shell and Tube Heat Exchanger with Graphene Oxide Nanofluid" discusses the addition of TiO_2 nanoparticles were added to conventional fluids such as water, ethylene glycol, and a blend of the two base fluids, which had lower heat conductivity than solid particles. Nanoparticles have a larger surface area than microparticles, which allows them to perform better in terms of heat transfer and stress stability. Increasing the rate of heat transfer in a double-pipe heat exchanger is one way to improve heat performance. The heat transfer characteristics decrease as the concentration of ethylene glycol increases, while the addition of TiO_2 nanoparticles improves the heat transfer characteristics due to the higher thermal conductivity.

Chapter 6 entitled "Microstructure Change of Aluminum 6061 on Natural and Artificial Aging" studies about the formation Mg_2Si due to artificial aging after natural aging that affected the microstructure of aluminum 6061. In this study, specimens before being treated with artificial aging with variations in holding time experienced natural aging for 7 days. The purpose of this study was to examine the changes in the microstructure of aluminum 6061 due to heat treatment. Previous research has shown that temperature and holding time in the heat treatment process have a very important role in changing the microstructure of the specimen. Changes in microstructure are caused by the formation of precipitates, where in aluminum 6061 the precipitate formed is Mg_2Si.

Chapter 7 entitled "Characterization of Self-Healing Concrete Incorporating Plastic Waste as Partial Material Substitution" discusses the effect of adding plastic waste on the characterization of self-healing concrete, such as compressive strength, tensile strength, flexural strength, and water permeability. Self-healing concrete can repair cracks created by internal and external work events, and it can also act as a corrosion-resistant barrier for reinforcement bars. By employing plastic waste and additive materials that can increase the performance of concrete, the collaboration of these materials contributes to the achievement of green chemistry for the future cement and building industry. Furthermore, the use of these materials has some implications in terms of the environment, production costs, and public policy.

Chapter 8 entitled "Graded Concrete: Towards Eco-friendly Construction by Material Optimisation" presents the findings on testings and simulation of graded concrete on the scope of material and its application as a building flexural element. The resulting compressive strength and modulus of elasticity of graded concrete have been studied on varied strength disparities. Further research looked into the flexural behavior of graded concrete beams in terms of load-deflection relationships and ductility, comparing them to conventional concrete beams with changing concrete strength in compression and tensile fibers, as well as varying reinforcing ratios. The evaluation of the prediction of building behavior utilizing graded concrete as beam components is also discussed. The findings suggest that optimizing cement in graded concrete technology does not degrade structural element performance. Graded concrete structural elements are more brittle; however, this can be solved by altering the reinforcing ratio of the members. Furthermore, the usage of graded concrete in building structures maintains the structural behavior that meets the performance control requirements outlined in design regulations.

Chapter 9 entitled "Performance of Surgical Blades from Biocompatible Bulk Metallic Glasses and Metallic Glass Thin Films for Sustainable Medical Devices Improvement" demonstrates the potential application of materials with the structure other than crystalline, amorphous structure, namely bulk metallic glass (BMG) and metallic glass thin film (MGTF) for surgical blade. A data set of biocompatible alloy systems had been developed in decades are provided in an informative way, followed by their respective material properties, that is, glass forming ability, mechanical and corrosion properties, and biocompatibility test reports. Turning to the amorphous materials conversion into surgical blade, several fabrication techniques are employed: wire cutting, grinding, and fine polishing for BMG blades; and sputtering for MGTF coating onto commercial blade substrate. The performance of the resulting amorphous-structured blades in terms of scratch performance, cutting ability, durability, as well as the biocompatibility investigation is discussed in the following part of this chapter. Finally, based on the material excellences, the final part brings up the capability of such materials to answer the medical challenges in the future.

Chapter 10 entitled "Synthesis and Characterization of Zinc Ferrite as Nanofluid Heat Exchanger Deploying Co-precipitation Method" discusses the effect of pH value on the synthesis of zinc ferrite using co-precipitation method. The pH value will affect the morphology (characterized by scanning electron microscope), phase identification (characterized by X-ray diffraction), molecular bonding (characterized by Fourier transform infrared), and also the performance of nanofluid in heat exchanger shell and tube. Nanofluid zinc ferrite will be a promising material to increase the efficiency of heat exchanger.

Chapter 11 entitled "A Study of Risk Assessment in the Nanomaterials Laboratory of Mechanical Engineering Department and the Materials Physics Laboratory of Department of Physics at State University of Malang" presents the data of risk assessment in nanomaterials laboratory in State University of Malang. The risk assessment was carried out by determining the exposure rating and hazard rating of each laboratory to obtain the risk level, based on risk matrix. Exposure rating is obtained from the duration of working with nanomaterials. The longer the laboratory users doing nanomaterials synthesis, the greater the potential of nanomaterials exposure to their body. Moreover, understanding of the condition of nanomaterials laboratory will minimize and prevent the risk of hazard. It will also enhance the safety of laboratory.

Sustainability advances ensure that technology becomes clean, cheap, high-efficiency energy, and environmentally friendly resources. Thereof, dye-sensitized solar cell (DSSC) has attracted to implementing sustainable energy sources. Chapter 12 entitled "Fabrication and Characterization of Dye-Sensitized Solar Cell in Various Metal Oxide Structure" presents the photovoltaic characteristics of TiO_2 and ZnO as photoanodes to develop sustainable energy sources in DSSC because those are non-toxic, biocompatible, chemically stable materials and have a high refractive index. In this chapter, the properties and photovoltaic performances of various components of DSSC will be shown from some characterization.

Chapter 13 entitled "Characterizations of Amino-Functionalized Metal-Organic Framework Loaded with Imidazole" studies the functionalization of amine and imidazolium into zirconium and iron-based metal-organic frameworks (MOF) as a preliminary study for the potential use of MOF as green solid-state electrolyte in lithium-ion battery. Several material characterization tools were used to identify the lattice structure, the surface morphology, and the main functional groups in these MOF. Intensification of crystal peaks from X-ray diffractograms was observed for functionalized MOF and the three-dimensional frameworks were retained, indicating a stable framework structure after incorporated with external molecules. The surface of the modified MOF was smooth, which is useful for the production of safe solid-state electrolyte.

Chapter 14 entitled "Green Removal of Bisphenol A (BPA) from Aqueous Media Using Zr-Based Metal-Organic Frameworks" presents the adsorption characteristics—kinetics, isotherm, and thermodynamics of the unique Zr-based MOFs in removing BPA. BPA is an endocrine-disrupting chemical in the flora and fauna ecosystem and is a great public health concern. On the other hand, MOFs have high specific surface area, high porosity, uniform pore volumes, and open active metal sites that offer great potential for adsorption demands in green approaches. This chapter of UiO-66 (Zr) MOF introduces the future recyclable adsorbent for the effective removal of BPA from aqueous media.

ACKNOWLEDGMENT

The authors extend their special thanks to the Faculty of Computing and Informatics, Universiti Malaysia Sabah, for the financial and computing facility supports that have made the completion of this book possible.

REFERENCE

[1]   https://sdgs.un.org/goals (Retrieved on 30 March 2022).

2 Phase Identification, Morphology, and Compressibility of Scallop Shell Powder (Amusium Pleuronectes) for Bone Implant Materials

Poppy Puspitasari[1]* and Diki Dwi Pramono[2]
[1] Faculty of Engineering, State University of Malang, Indonesia;
Center of Advanced Materials and Renewable Energy, State
University of Malang, Indonesia
[2] Faculty of Engineering, State University of Malang, Indonesia
*Corresponding author: Poppy Puspitasari, poppy@um.ac.id

CONTENTS

DOI: 10.1201/9781003320746-2

2.1   INTRODUCTION

Scallops with the scientific name Amusium pleuronectes contain calcium in the form of calcium carbonate ($CaCO_3$) as the main constituent of scallop shells, which is as much as 97–99% (Raju et al., 2003). Scallops are one of the marine biota that has a fairly wide distribution. The distribution of scallops starts in the South China Sea, the Indian Sea, Indo-China, Japan, the Philippines, Papua New Guinea, Indonesia and Australia (Carpenter & Niem, 1998). The scallop-producing area in Indonesia is the northern coast of Java, Central Java province. Scallop catches in January–March 2008 reached more than 41 tons, and it is estimated that the waste reached 21.73 tons (Widowati et al., 2008).

The existence of scallop waste requires special attention, and so far the processing of scallop waste in Indonesia is the most widely used for the manufacture of accessories (Mufidun et al., 2016). Meanwhile, in research, scallop shells are used as a biodiesel catalyst, used for the manufacture of fillers as the basic material for making composite boards, and used for the manufacture of chitosan as a coagulant for wastewater (Buasri et al., 2014; Mufidun et al., 2016; Pradifan et al., 2016). Scallop shells can also be used as a basic material for hydroxyapatite which is then used for bone implant materials (Syafaat & Yusuf, 2019). The scallop shell also contains a type of calcium phosphate bioceramic that allows it to be used as a material for bone implants (Rachman et al., 2019). The high levels of calcium present in scallop shells provide a potential opportunity for scallop shells to replace the role of synthetic calcium and calcium from beef bones, which are generally more expensive in terms of economy and more difficult to supply.

The manufacture of scallop shell powder can be done with a synthesis process, both top-down and bottom-up approaches. The bottom-up approach includes sol-gel, precipitation, sputtering and spin coating, while the top-down approach includes milling, lithography, and electrodeposition methods. Among the milling methods, the ball milling method is the simplest method and the time used is also relatively fast. Ball milling is able to produce smaller particles in a shorter milling time (Puspitasari, 2017).

2.1.1   PREVIOUS STUDIES

Several previous studies undertaken as theoretical studies carried out by several experts are detailed in Table 2.1.

2.1.2   SCALLOPS

The scientific name of scallop shell is Amusium pleuronectes and it is a member of the Pectinidae family. There are more than 30 genera and about 350 species in the Pectinidae family. The habitat of the scallops is in the bottom sea area with a tropical climate (Swennen, 2001). Scallop shells can live in almost all marine waters in the world, and therefore they are dubbed as cosmopolitan shells. The distribution of scallops starts from the eastern Indian Ocean and western Pacific, Myanmar, Indonesia, Papua New Guinea, northern Taiwan, the province of China, southern

TABLE 2.1
Previous Studies

Author	Results
Agustini et al. (2011)	Utilizing waste scallop shells (Amusium pleuronectes) in the manufacture of calcium-rich cookies. The separation of calcium from scallop shells is carried out by deproteinase, namely, by removing protein from the shell by means of protein hydrolysis. Extraction of calcium from scallop shells is done by protein hydrolysis process using hydrochloric acid (HCl) solution. The results showed that increasing the concentration of scallop shell flour had a very significant effect ($\alpha = 0.01$) on water content, ash, fat, protein, calcium, phosphorus, and cookie hardness. The addition of simping shell flour with a concentration of 7.5% produced cookies with the highest calcium content (6.57%), phosphorus content (1.58%), ash (6.95%), carbohydrates (52.31%), and cookie hardness value (1.06 KgF).
Buasri et al. (2014)	Utilizing scallop shell waste as a catalyst to make biodiesel. The scallop shell waste was calcined at 1000°C for 4 hours and the catalyst characterization was carried out by XRD, XRF, SEM, and BET surface area measurements. Under optimal reaction conditions of 10% by weight of catalyst, 9:1 methanol/oil molar ratio and at 65°C, FAME conversion was 95.44% and was achieved in 3 hours. It was found that the catalyst from shellfish waste shows high catalytic activity and environmentally friendly properties, has a potential opportunity to be used in the biodiesel production process as a heterogeneous base catalyst.
Hariharan et al. (2014)	Made calcium carbonate nanoparticles synthesized using the precipitation method from clam shells using chitosan as a precursor. The synthesized calcium carbonate was characterized using SEM, XRD, UV-Vis, and FTIR spectroscopy. The results were compared with commercial calcium carbonate nanoparticles. Shellfish shells are a potential source of calcium carbonate. The ingredients used are natural and also from by-products of the seafood industry.
Pornchai et al. (2016)	Conducting research to make CaO from eggshell industrial waste. Eggshell waste is calcined at a temperature of 800°C with a temperature variation of 1 to 4 hours, then the physical and chemical properties of calcium oxide resulting from calcined eggshell waste are systematically characterized by XRF, XRD, SEM, particle size analyzer, and gas adsorption experiments. The results showed that in terms of energy saving, calcination of eggshell industrial waste at a temperature of 800°C for 1 hour was sufficient to produce high-purity calcium oxide. In addition, calcium oxide with a high surface area can also be obtained at optimal calcination temperature and time. This study also shows that the physical and chemical properties of calcium oxide obtained from eggshell waste are comparable to those of commercial calcium oxide.
Mohagheghian et al. (2018)	Shellfish-Fe_3O_4 nanoparticles were synthesized by co-precipitation method under vacuum conditions. The efficient coating of Fe_3O_4 nanoparticles into scallop shells was identified by FT-IR, XRD, SEM, EDX, and VSM analysis. The removal efficiency of AR14 by scallop shell-Fe_3O_4 nanoparticles was greater than that by scallop shell alone and Fe_3O_4 alone. Maximum adsorption was observed under acidic conditions. The removal efficiency of AR14 increased with increasing adsorbent dose, but decreased with increasing concentration and initial temperature of AR14.

(continued)

TABLE 2.1 (Continued)
Previous Studies

Author	Results
	The adsorption capacity of AR14 on the adsorbent was slightly influenced by the type of ionic strength except for carbonate ions. In the kinetic study, the rate of removal was better described by the pseudo-second-order model than the pseudo-order model and the intra-particle diffusion model. The adsorption isotherm was analyzed using the Langmuir and Freundlich equations. The experimental results reveal that the adsorption reaction is an exothermic process. The absorption efficiency of AR14 by scallop shell-Fe_3O_4 nanoparticles is maintained even after six consecutive cycles.
Puspitasari et al. (2019)	Making $CaCO_3$ and CaO from chicken egg shells using a ball milling process with variations in sintering time of 60, 90, and 120 minutes, with a milling time of 10 hours. Then the samples were characterized using SEM-EDX and XRD. The results showed that the sample with a sintering time of 90 minutes had the highest degree of crystallinity. Samples with a sintering time of 60 minutes had the smallest grain size of 14.6 nm, and the largest of 32.2 nm. Eggshell nano powder sintered for 60 minutes had the lowest element content, while that sintered for 120 minutes had the highest element content.
Syafaat and Yusuf (2019)	Making hydroxyapatite (HAp) from scallop shells using the precipitation method and varying the concentration of Ca/P. Scallop shells were calcined at 1000°C to produce CaO. The precursor was a suspension of $Ca(OH)_2$ and a solution of H_3PO_4, and HAp was synthesized by mixing the two precursors and calcining at 1050°C. The purpose of this study was to investigate the effect of Ca/P concentration. Samples were characterized by XRD, SEM-EDX, and FTIR. The XRD pattern showed an increase in crystallinity with increasing Ca/P concentration ratio. The lattice parameters and sample density values of 0.5/0.3 are close to the theoretical values. The crystal size of the HAp powder ranged from 79.750 ± 0.066 to 90.932 ± 0.071 nm. Ca/P ratios of 0.5/0.3 and 1/0.6 were 1.68 and 1.67, respectively. The FTIR spectra confirmed the presence of the PO_4^{3-} and OH^- functional groups.
Rachman et al. (2019)	Making lime precursors to form calcium phosphate-type bioceramics from scallop shells. Scallop shells were synthesized using the precipitation method with pH 6–7, 7–8 and 8–9, then calcined at temperatures of 800°C and 900°C. The results were then characterized by XRD which showed that the calcination temperature did not have a significant effect on the crystal form, at pH 6–7 and pH 7–8 β-tricalcium phosphate minerals were formed, while at pH 8–9 hydroxyapatite minerals were formed. FTIR test results at pH 7–8 showed that at wavelength 3650–3000 cm-1 there was no OH vibration, whereas at pH 8–9 and a calcination temperature of 900°C there were typical peaks at wavelengths of 555.50 cm^{-1} and 609.51 cm^{-1} and OH vibrations occur at the wavelength 3650–3000 cm^{-1}, it indicates that hydroxyapatite formation occurs.

Japan, and southern Queensland (Habe, 1964). In Indonesia, the largest scallop-producing area is the northern coast of Java, Central Java province. Scallop catching in January–March 2008 reached more than 41 tons, and it is estimated that the waste reached 21.73 tons (Widowati et al., 2008)

The scallop shell contains calcium in the form of calcium carbonate ($CaCO_3$) as the main composition, which is 97–99% (Raju et al., 2003). The nacre layer on the scallop shell contains calcite compound ($CaCO_3$) which has a multi-layered micro-structure, is isotropic, and has a higher hardness value than other natural calcite crystals (Li & Ortiz, 2014).

2.1.3 CALCIUM CARBONATE ($CaCO_3$)

Calcium carbonate is an inexpensive commercially available inorganic mineral (Noviyanti et al., 2015). In nature, $CaCO_3$ can be found in limestone in various forms, such as marble and limestone (Patnaik, 2001). Limestone is a natural rock that is commonly found in Indonesia. It is a solid rock that contains a lot of calcium carbonate (Lukman et al., 2012).

Calcium carbonate is not only derived from limestone, but it is also the main component of shells of marine organisms, snails, egg shells, and pearls (Bahanan, 2010). But to get $CaCO_3$ from these materials, you must first carry out a process called synthesis. There are several methods used to synthesize $CaCO_3$ from natural materials, including the ball milling method, the sol-gel method, the hydrothermal method, the sono-chemical method, the precipitation method, the solids method and the microwave method.

Calcium carbonate is generally white in color and is often found in limestone. It consists of several elements, namely, calcium (Ca), carbon (C), and oxygen (O). Each element is strongly bonded to three oxygens, and the bonds are much looser than the bonds between carbon and calcium in a single compound (Bahanan, 2010).

Calcium carbonate has several physical properties, including its solubility in water (15 mg/L at 25°C) and in dilute minerals (Patnaik, 2001). Calcium carbonate has three polymorphs, namely, calcite, aragonite, and vaterite. Calcite decomposes at 825°C, while aragonite decomposes at 1339°C under a pressure of 102.5 atm. Calcite has a density value of 2.71 g/cm^3, while aragonite has a density value of 2.83 g/cm^3. However, calcite and aragonite are water-soluble materials (Patnaik, 2001). Table 2.2 presents the characteristics of calcium carbonate. Three polymorphs of calcium carbonate—aragonite, calcite, and vaterite—are found in numerous biominerals and biocomposites. Aragonite, for example, forms naturally in almost all mollusk shells, and as the calcareous endoskeleton of corals.

2.1.4 BALL MILLING

Ball milling is a method of synthesizing nanomaterials that utilizes the energy of the collision between the crushing balls and the walls of the container (Puspitasari, 2017). At this time an innovation has been made on the ball milling machine (Figure 2.1), by changing the mill rotation to a planetary ball mill in the container

TABLE 2.2
Characteristics of Calcium Carbonate

IUPAC Name	Calcium Carbonate
Chemical formula	$CaCO_3$
Shape	Congested
Color	White/colorless
Density	2.8 g/cm³
pH	9.5–10.5 on 100 g/L 20° C (porridge)
Melting point	825° C (decomposition)
Decomposition temperature	825° C
Solubility in water	0.014 g/L on 20° C
Density	400–1400 kg/m³

Source: Yasue and Arai, 1995.

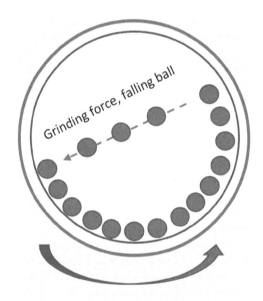

Rotating Milling Bowl

FIGURE 2.1   Schematic presentation of ball milling.

that has levers on both sides. They are use for adjusting the optimal rotation angle, and also during the ball mill process. The material is stabilized using a chemical solution, such as polyvinyl alcohol (PVA), or can use polythilene glycol (PEG) to form stable nanocolloids (Puspitasari, 2017).

Ball milling consists of two main components, namely, the ball and the vial. At the time of manufacture of nanomaterials, the material is inserted into the tube along with the crushing ball. Then the ball milling is driven by rotation or high-frequency vibration, and as a result of this ball milling motion the trapped material between the crushing ball and the tube wall will collide with each other, thereby causing the material to deform. The deformation of the material causes the fragmentation of the structure, thereby making it break up into smaller structures (Puspitasari, 2017). There are several factors that affect the results of this ball milling process, such as time, speed, temperature, crushing ball size, pressure, process control agent (PCA) percentage, and powder weight composition (Suryanarayana, 2004).

2.1.5 SINTERING

Sintering is a heat treatment process on powders with a high temperature below their melting point, the purpose of which is to obtain structural changes in the material, such as grain growth, reduction in the number and size of porosity, increased density, and volume shrinkage of the material (Derlet, 2017; Kang, 2005). During the sintering process, the particles unite so that their density increases (agglomeration), and grain boundaries are formed, which are the initial stage of recrystalization and besides that there will also be vaporized gas (Budihartomo, 2012).

There are several factors that can affect the results of the sintering process, namely, temperature, holding time, pressure, atmosphere, heating speed, and cooling speed (Kang, 2005). The higher the sintering temperature or the longer the holding time used, the higher the sinter density, the smaller the porosity, the higher the compressive strength and the modulus of elasticity (Safrudin & Widyastuti, 2012).

2.2 MATERIALS AND METHODS

The base material, scallop shell, that was pounded/crushed before was divided into two 300 g each and was put into the vial ball mill. Acetone was added into the ball mill for 100 ml each. Milling duration was five hours, with a speed rotation of 400 rpm. The milled samples then were sintered at 1100°C for 60, 90, and 120 minutes. The sintered samples then crushed to avoid agglomerate. The $CaCO_3$ synthesis results then were characterized by using XRD PANalytical X'Pect (λ=1.54 Å) to find the phase and crystal size. Morphology and elements characterization were found using the SEM-EDX Inspect-S50 type, whereas the functional groups were characterized by using the FTIR DLATGS, InGaAs type. Compressibility of the sample was tested with Universal Testing Machine type IL-904.

2.3 RESULTS AND DISCUSSION

2.3.1 PHASE CHARACTERIZATION

Phase identification was carried out by looking at the results of XRD characterization to compare the crystalline phases in scallop shell powder and to analyze grain size, crystal orientation, phase structure, and crystal defects in each phase (Bunaciu et al., 2015). To calculate the crystallite size of the scallop shell powder, Scherrer equation was used, as shown in Equation 2.1 (Puspitasari et al., 2019):

TABLE 2.3
Intensity, FWHM, D-spacing, and Crystal Size of Scallop Shell Powder

	XRD *Peak*			
Sample	**Intensity** (*counts*)	**FWHM** (rad)	**D-spacing** (Å)	**Crystal Size** (nm)
CKS-NS	1400	0.002061234	3.0352	69.55
CKS-60	712	0.001373574	2.40372	106.57
CKS-90	1041	0.001373574	2.40367	106.58
CKS-120	889	0.001373574	2.40163	106.59

$$d = \frac{k\lambda}{\beta\cos\theta}$$
2.1

where d is the diameter of the crystal, β is the full-width half maximum (FWHM), K is the constant (0.89; 0.9), and λ is the wavelength (1,5406 Å). The results of the calculation of the XRD data using the above formula are shown in Table 2.3.

Table 2.3 shows that there is a significant difference in crystal size in the scallop shell powder samples without sintering (CKS-NS) and with the sintering process. The crystal size of the sample without the sintering process (CKS-NS) was 69.55 nm, and the sample with the sintering temperature of 1100°C for 60 minutes (CKS-60), 90 minutes (CKS-90), and 120 minutes (CKS-120) was 106.57, 106.58 and 106.59 nm, respectively.

From Figure 2.2, it can be seen that the XRD graph shows that there is a difference between the scallop shell powder sample without sintering and with the sintering process. In the sample of scallop shell powder without sintering (CKS-NS), the highest peak is at $2\theta = 29.4285°$ and at this peak shows the diffraction peak of the rhombhohedral $CaCO_3$ phase (Tangboriboon, et al., 2012). The sample of scallop shell powder with a sintering process temperature of 1100°C with time variations of 60 minutes (CKS-60), 90 minutes (CKS-90), and 120 minutes (CKS-120), the highest peak was at $2\theta = 37.4137°$, $2\theta = 37.4145°$, and $2\theta = 37.4476°$ at the three peaks contained CaO content. From the XRD graph, the scallop shell powder sample with the sintering process has no peak at $2\theta = 29.4285°$, and this indicates that $CaCO_3$ is transformed into CaO (Mosaddegh & Hassankhani, 2014). The sintering process with a temperature of 1100°C on the scallop shell powder samples resulted in $CaCO_3$ being decomposed into CaO. Decomposition of $CaCO_3$ into CaO can occur if the minimum sintering temperature is 954°C (Rujitanapanich et al., 2014). On the other hand, $Ca(OH)_2$ peaks were also formed in the scallop shell powder samples with sintering processes of 60, 90, and 120 minutes. The formation of $Ca(OH)_2$ is caused by the reaction between CaO and H_2O in the air (Nurhayati et al., 2016). The $Ca(OH)_2$ phase comes from water molecules that are adsorbed on the surface of CaO, where CaO is known to be hygroscopic, so it is very easy to absorb water vapor from the air (Mohadi et al., 2013). The percentage of $Ca(OH)_2$ phase contained in the scallop

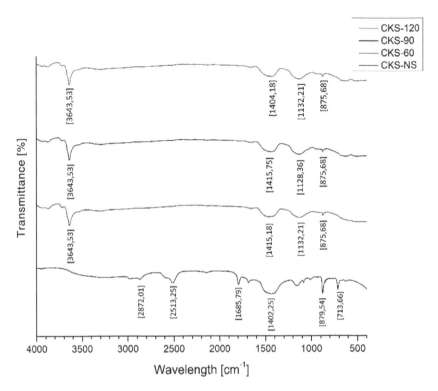

FIGURE 2.2 XRD graph of scallop shell powder.

shell powder samples was analyzed using MATCH software; the results of the analysis showed that the samples were sintered for 60 minutes (CKS-60), 90 minutes (CKS-90), and 120 minutes (CKS-120) by 13.5%, 10.1%, and 6.4%. From the results of the analysis using the MATCH software, the crystal form of the scallop shell powder sample is also obtained, where the sample without the sintering process has a trigonal crystal shape (hexagonal axes) and after sintering the crystal shape changes to cubic from (Supriyanto et al., 2019).

2.3.2 MORPHOLOGICAL CHARACTERIZATION

The morphology of the scallop shell powder samples was observed using scanning electron microscope (SEM). Morphological differences between samples without sintering and those with sintering are shown in Figure 2.3.

SEM test results for scallop shell powder samples without sintering process (CKS-NS; Figure 2.3(a)) have irregular particle shape due to agglomeration during deposition after milling process (Supriyanto et al., 2019; Ahmad et al., 2015). After the sintering process, the shape of the particle structure of the scallop shell powder sample changes to an interconnected skeleton (Witoon, 2011). The results of the grain size analysis in the SEM image using ImageJ Software showed that the scallop shell powder sample (CKS-NS) without sintering has an average grain size of

FIGURE 2.3 Morphology of scallop shell powder: (a) CKS-NS, (b) CKS-60, (c) CKS-90, and (d) CKS-120.

551.44 nm, while the scallop shell powder sample with a sintering process of 60 minutes (CKS-60), 90 minutes (CKS-90), and 120 minutes (CKS-120) showed a grain size of 603.68, 619.39, and 631.17 nm, respectively. This shows that the sintering process causes the phenomenon of grain growth (Carter & Norton, 2007).

Figure 2.4 and Table 2.4 show the percentage of porosity in the scallop shell powder sample obtained from the analysis of the SEM image of the scallop shell powder sample at magnification 5000×, 10000×, 25000×, 50000×, and 100000× using OriginPro 2016 software.

Figure 2.4 shows that the scallop shell powder sample without the sintering process (CKS-NS) has a higher percentage of porosity compared to the sintered scallop shell powder sample (CKS-60, CKS-90, and CKS-120). This is because in the sintering process there is a reduction in the number and size of the porosity (Derlet, 2017).

The sample with a 60-minute sintering process (CKS-60) has a lower percentage value of porosity compared to the scallop shell powder sample with 90 minutes (CKS-90) and 120 minutes (CKS-120) sintering. This is probably due to the release of CO_2 gas during the sintering process, which causes pores in the sample with the sintering

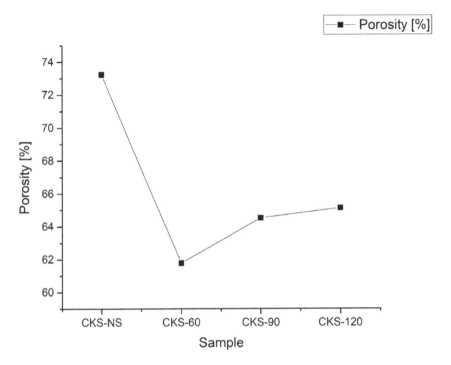

FIGURE 2.4   Graph of porosity percentage in scallop shell powder samples.

TABLE 2.4
Percentage of Porosity in Scallop Shell
Powder Samples

Sample	Average Porosity (%)
CKS-NS	73.23
CKS-60	61.78
CKS-90	64.51
CKS-120	65.11

process (Mosaddegh & Hassankhani, 2014; Witoon, 2011). The longer the sintering process, the more the release of the CO_2 gas and it causes the samples with a sintering time of 90 and 120 minutes to have a greater percentage of porosity.

2.3.3   FOURIER-TRANSFORM INFRARED SPECTROSCOPY (FTIR) ANALYSIS

Identification of the distribution of functional groups in the scallop shell powder samples was carried out using FTIR. Figure 2.5 presents a graph of the FTIR results on a sample of scallop shell powder.

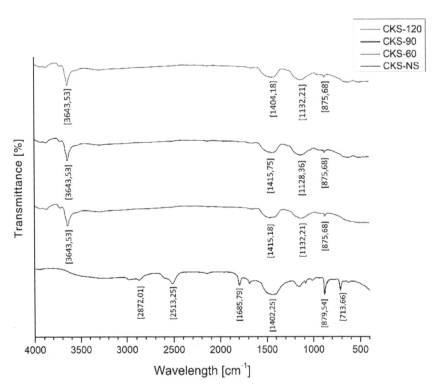

FIGURE 2.5   FTIR graph of scallop shell powder samples.

TABLE 2.5
Distribution of Functional Groups Contained in Scallop Shell Powder Samples without Sintering Process (CKS-NS)

Wavelength Peak (cm⁻¹)	Information
713.66; 879.54; 1402.,25	Carbonate groups and natural characteristics of dolomite
1012.63	*Stretching* C–O
1685.79; 2513.25; 2872.01	$CaCO_3^{-2}$ band combination

Figure 2.5 shows the molecular bond of scallop shell samples without sintering (CKS-NS) and Table 2.5 describes the distribution of molecular bond of scallop shell without sintering according to the peak. There is a sharp peak at wavelengths of 713.66 cm⁻¹, 879.54 cm⁻¹, and 1402.25 cm⁻¹; it shows the carbonate group and the characteristics natural dolomite contained in the sample (Khiri et al., 2016; Correia et al., 2017). At wavelength 1012.63 cm⁻¹, stretching of C–O occurs (Kamalanathan et al., 2014). In addition, small infrared absorption spectra shown at wavelengths of 1685.79 cm⁻¹, 2513.25 cm⁻¹, and 2872.01 cm⁻¹ were caused by different combinations of $CaCO_3^{-2}$ bands (Kamalanathan et al., 2014; Gunasekaran et al., 2006).

TABLE 2.6

Distribution of Functional Groups Contained in Scallop Shell Powder Samples with Sintering Process of 60 Minutes (CKS-60), 90 Minutes (CKS-90), and 120 Minutes (CKS-120)

Wavelength Peak (cm⁻¹)	Information
3643.53	OH band stretch
1415.18 and 1132.21	Asymmetric C–O with vibration
875.68	Ca–O

From the FTIR graph for the scallop shell powder samples with the sintering process at temperatures of 1100°C for 60 minutes (CKS-60), 90 minutes (CKS-90), and 120 minutes (CKS-120), there was no significant difference. Table 2.6 describes that at the peak of 3643.53 cm⁻¹, it shows that there is a stretching of the OH bond during water adsorption by CaO (Ahmad et al., 2015). The bands 1415.18 cm⁻¹, 1132.21 cm⁻¹, and 875.68 cm⁻¹ are associated with C–O asymmetry with vibrations of the carbonate group (Tizo et al., 2018; Doostmohammadi et al., 2011). At the peak of 875.68 cm⁻¹ it shows a Ca–O bond (Tangboriboon et al., 2012).

2.3.4 IDENTIFICATION OF THE COMPRESSIBILITY OF SCALLOP SHELL POWDER SAMPLES

The compressibility test was carried out to determine the value of tapped density and compressibility of the scallop shell powder sample.

Figures 2.6 and 2.7 present the results of compressibility testing of scallop shell powder samples with variations in loading of 1000 and 2000 KgF. From the figure, it can be seen that the scallop shell powder sample without the sintering process (CKS-NS) has a tapped density and compressibility at loadings of 1000 and 2000 KgF, and these values are lower when compared to the sample of scallop shell powder with a sintering process (CKS- 60, CKS-90, and CKS-120).During the sintering process the porosity of the sample will decrease, which will result in an increase in density (Beddoes & Bibby, 1999). Apart from being caused by porosity, compressibility values are also affected by grain shape; the amount of irregular grain shape of the sample will cause a decrease in compressibility value (Schatt & Wieters, 1997). In the SEM results on samples of scallop shell powder without a sintering process (CKS-NS), the grain size looks irregular, causing the compressibility value of samples of scallop shell powder without a sintering process to be smaller than those with a sintering process.

The sample with a sintering time of 60 minutes (CKS-60) has a higher value of tapped density and compressibility than the sample with a sintering time of 90 minutes (CKS-90) and 120 minutes (CKS-120). From the results of SEM image analysis using ImageJ software, it is shown that if the holding time during the sintering process is longer, the average grain size increases, and increasing grain size causes the compressibility value to decrease (Espin et al., 2019). Figure 2.3 shows the sample

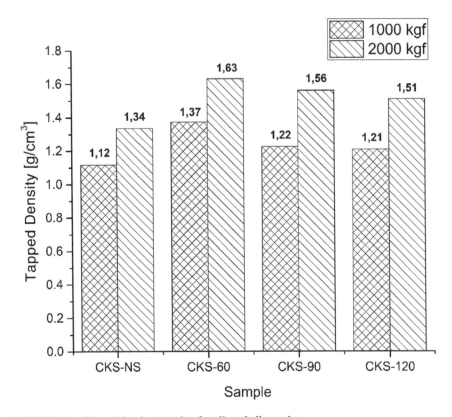

FIGURE 2.6   Tapped density sample of scallop shell powder.

with 60-minute sintering process (CKS-60) has the smallest porosity value, which suggests that when the porosity value is smaller, the compressibility value and compression ratio value will increase (Xu et al., 2017).

2.3.5   RELATIONSHIP BETWEEN COMPRESSIBILITY, GRAIN SIZE, AND POROSITY

Figure 2.8 and Table 2.7 show the relationship between porosity and compressibility in scallop shell powder samples. From the figure, it can be seen that the scallop shell powder sample without a sintering process (CKS-NS) has the lowest compressibility value and highest porosity percentage value, while the scallop shell powder sample with a sintering process of 60 minutes (CKS-60) has the highest compressibility value and lowest porosity percentage value. It can be concluded that the compressibility value is inversely proportional to the percentage value of porosity. This is in accordance with research conducted by Xu et al. (2017) that if the porosity value of a sample is higher, the compressibility value will decrease.

Figure 2.9 and Table 2.8 show the relationship between the average value of grain size and the compressibility value of the scallop shell powder sample. Figure 2.10 and Table 2.9 show that the average value of the grain size of the scallop shell powder

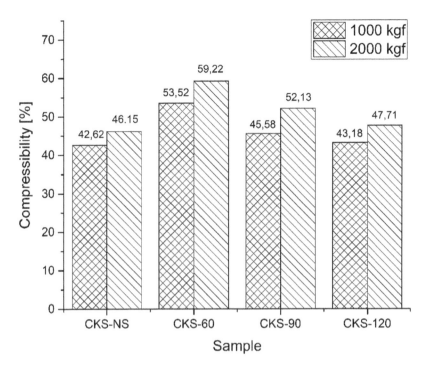

FIGURE 2.7 Compressibility of scallop shell powder samples.

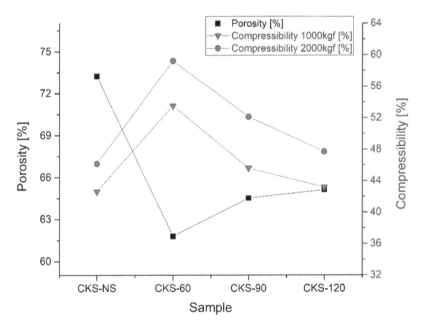

FIGURE 2.8 Graph of the relationship of porosity and compressibility in scallop shell samples.

TABLE 2.7
Porosity and Compressibility Values of Scallop Shell Powder Samples

Sample	Porosity (%)	Compressibility 1000 KgF (%)	Compressibility 2000 KgF (%)
CKS-NS	73.23	42.62	46.15
CKS-60	61.78	53.52	59.22
CKS-90	64.51	45.58	52.12
CKS-120	65.11	43.18	47.70

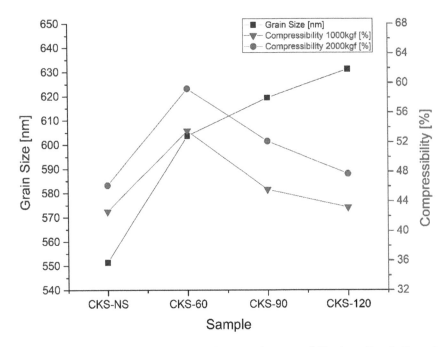

FIGURE 2.9   The relationship between grain size and compressibility in scallop shell powder samples.

TABLE 2.8
Value of Grain Size and Compressibility of Scallop Shell Powder Samples

Sample	Grain Size (nm)	Compressibility 1000 KgF (%)	Compressibility 2000 KgF (%)
CKS-NS	551.44	42.62	46.15
CKS-60	603.68	53.52	59.22
CKS-90	619.39	45.58	52.12
CKS-120	631.17	43.18	47.70

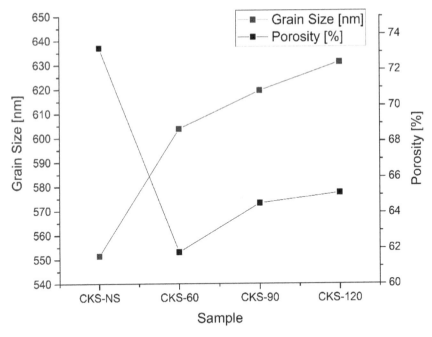

FIGURE 2.10 Relationship between grain size and porosity in scallop shell powder samples.

TABLE 2.9
Porosity Value and Grain Size of Scallop Shell Powder Samples

Sample	Grain Size (nm)	Porosity (%)
CKS-NS	551.44	73.23
CKS-60	603.68	61.78
CKS-90	619.39	64.51
CKS-120	631.17	65.11

sample with the sintering process increases with longer sintering process and the compressibility value of the scallop shell powder sample with the sintering process decreases with the longer sintering time. Based on the opinion expressed by Asthana et al. (2006), the grain size of the sample affects the compressibility value of the sample. This is because the longer the holding time in the sintering process causes grain growth, the larger the grain size and the lower the compressibility value (Espin et al., 2019). The compressibility value of the sample is also influenced by the grain shape. Schatt and Wieters (1997) contend that the more irregular the grain shape of the sample, the more it will decrease the compressibility value.

Changes that occur due to the sintering process include changes in grain size, grain shape, and also pore size. The number of pores formed also depends on the particle size distribution and the shape of the grains based on the synthesis process.

To improve the mechanical properties such as compressibility and thermal conductivity, it is necessary to eliminate the number of pores by the sintering process. In the sample of scallop shell powder without a sintering process (CKS-NS), the crystal size and grain size automatically become small so that there is a lot of porosity, while in the samples with a sintering process (CKS-60, CSK-90, and CKS-120) the grain size increases and therefore the porosity decreases. In the sintering process with variations in the length of sintering time, it shows that the phenomenon of grain growth and grain shrinkage occurs (Carter & Norton, 2007).

2.4   CONCLUSION

1. Phase identification results show that there is a phase change from $CaCO_3$ to CaO in samples with a good sintering process with a time of 60, 90, and 120 minutes; in samples with the sintering process there is also a $Ca(OH)_2$ phase. In samples without a sintering process, the crystal size was 69.55 nm, and in samples with a sintering time of 60, 90, and 120 minutes, the crystal size was 106.57, 106.58, and 106.59 nm, respectively.
2. The results of morphological identification showed that the morphology of the scallop shell powder sample without the sintering process had an irregular shape, while those with the sintering process had an interconnected skeleton shape.
3. The results of the identification of the distribution of functional groups showed that samples without the sintering process contained the $CaCO_3$ functional group, while those with the sintering process contained the CaO and $Ca(OH)_2$ functional groups.
4. The results of the identification of compressibility with loading variations of 1000 and 2000 KgF showed that the sample of scallop shell powder without a sintering process (CKS-NS) had compressibility values and compressibility ratios, while samples with a sintering time of 60 minutes (CKS-60) had a value of tapped density and highest compressibility among other samples.
5. The best sample is a sample with a sintering time variation of 60 minutes (CKS-60) because it has a CaO phase, smaller porosity value, smaller average grain size, and greater compressibility when compared to other samples, which makes it possible to be applied as bone implant material.

REFERENCES

A. A. Bunaciu, E. gabriela Udriştioiu, and H. Y. Aboul-Enein, "X-Ray diffraction: instrumentation and applications," *Crit. Rev. Anal. Chem.*, vol. 45, no. 4, pp. 289–299, 2015, doi: 10.1080/10408347.2014.949616.

A. Buasri, P. Worawanitchaphong, S. Trongyong, and V. Loryuenyong, "Utilization of scallop waste shell for biodiesel production from palm oil – Optimization using Taguchi method," *APCBEE Procedia*, vol. 8, no. CAAS 2013, pp. 216–221, 2014, doi: 10.1016/j.apcbee.2014.03.030.

A. Doostmohammadi, A. Monshi, R. Salehi, M. H. Fathi, Z. Golniya, and A. U. Daniels, "Bioactive glass nanoparticles with negative zeta potential," *Ceram. Int.*, vol. 37, no. 7, pp. 2311–2316, 2011, doi: 10.1016/j.ceramint.2011.03.026.

A. Mohagheghian, R. Vahidi-Kolur, M. Pourmohseni, J. K. Yang, and M. Shirzad-Siboni, "Preparation and characterization of scallop shell coated with Fe_3O_4 nanoparticles for the removal of azo dye: Kinetic, equilibrium and thermodynamic studies," *Indian J. Chem. Technol.*, vol. 25, no. 1, pp. 40–50, 2018.

A. Mufidun and A. Abtokhi, "Pemanfaatan Filler Serbuk Cangkang Kerang Simping (Placuna placenta) Dan Matriks Poliester Sebagai Bahan Dasar Pembuatan Papan Komposit," *J. Neutrino*, vol. 9, no. 1, pp. 1–7, 2016.

A. Pradifan, E. Sutrisno, and M. Hadiwidodo, "Studi Penggunaan Kitosan Dari Limbah Cangkang Kerang Simping (Amusium Pleuronectes) Ssebagai Biokoagulan Untuk Menurunkan Kadar COD Dan TSS," *J. Tek. Lingkung.*, vol. 5, no. 3, pp 1–8, 2016.

A. Rachman, N. Sofiyaningsih, and K. Wahyudi, "Karakteristik Mineralogi Material Biokeramik Jenis Kalsium Fosfat Dari Cangkang Kerang Simping," *Keramik dan Gelas Indones.*, vol. 27, no. 2, pp. 77–93, 2019.

C. B. Carter and M. G. Norton, *Ceramic Materials: Science and Engineering*. New York: Springer Science, 2007.

C. H. L. Raju, K. V. Narasimhulu, N. O. Gopal, J. L. Rao, and B. C. V. Reddy, "Structural studies of marine exoskeletons: Redox mechanisms observed in the Cu-supported $CaCO_3$ surfaces studied by EPR," *Spectrochim. Acta – Part A Mol. Biomol. Spectrosc.*, vol. 59, no. 13, pp. 2955–2965, 2003, doi: 10.1016/S1386-1425(03)00122-7.

C. R. D. Swennen, R. G. Moolenbeck, N. Ruttanadakul, H. Hobbelink, H. Dekker, and S. Hajisamac, "The molluscs of the southern Gulf of Thailand," *Thai Stud. Biodivers.*, vol. 4, pp. 141–148, 2001.

C. Suryanarayana, *Mechanical Alloying and Milling*. New York: Marcel Dekker. 2004.

E. Mosaddegh and A. Hassankhani, "Preparation and characterization of nano-CaO based on eggshell waste: novel and green catalytic approach to highly efficient synthesis of pyrano [4, 3-b]pyrans," *Cuihua Xuebao/Chinese J. Catal.*, vol. 35, no. 3, pp. 351–356, 2014, doi: 10.1016/s1872-2067(12)60755-4.

F. Y. Syafaat and Y. Yusuf, "Influence of ca/p concentration on hydroxyapatite (Hap) from Asian moon scallop shell (Amusium pleuronectes)," *Int. J. Nanoelectron. Mater.*, vol. 12, no. 3, pp. 357–362, 2019.

I. Widowati, J. Suprijanto, I. Susilowati, T. W. Agustini, and A. B. Raharjo, "Small-scale fisheries of the Asian Moon Scallop Amusium pleuronectes in the Brebes Coast, Central Java, Indonesia". *International Conference on Environmental System*, Canada, CM 2008/ K:08. 2008.

J. Beddoes and M. J. Bibby. "Powder metallurgy,". In: J. Beddoes and M. J. Bibby, Editors. *Principles of Metal Manufacturing Processes*. Canada: Elsevier Butterworth-Heinemann, 1999, pp. 173–189.

J. Noviyanti and E. H. Sujiono, "Karakterisasi Kalsium Karbonat ($Ca(CO_3)$) Dari Batu Kapur Kelurahan Tellu Limpoe Kecamatan Suppa," *J. Sains dan Pendidik. Fis.*, vol. Jilid 11 N, pp. 169–172, 2015.

K. E. Carpenter and V. H. Niem, "The living marine resources of the Western Central Pacific. Volume 1: Seaweeds, corals, bivalves and gastropods" In*: FAO Species Identification Guide for Fishery Purposes*. Rome: Food and Agriculture Organization of The United Nations, 1998.

L. Li and C. Ortiz, "Pervasive nanoscale deformation twinning as a catalyst for efficient energy dissipation in a bioceramic armour," *Nature Materials*, vol. 13, no. May, pp. 1–7, 2014, doi: 10.1038/NMAT3920.

L. M. Correia, J. A. Cecilia, E. Rodríguez-Castellón, C. L. Cavalcante, and R. S. Vieira, "Relevance of the physicochemical properties of calcined quail eggshell (CaO) as a catalyst for biodiesel production," *J. Chem.*, vol. 7, no. 8590, pp. 1–13. 2017, doi: 10.1155/2017/5679512.

M. Hariharan *et al.*, "Synthesis and characterisation of $CaCO_3$ (Calcite) nano particles from cockle shells using chitosan as precursor," *Int. J. Sci. Res. Publ.*, vol. 4, no. 10, pp. 1–5, 2014, [Online]. Available: www.ijsrp.org.

M. J. Espin, J. M. P. Ebri, and J. M. Valverde, "Tensile strength and compressibility of fine $CaCO_3$ powders. Effect of nanosilica addition," *Chem. Eng. J.*, vol. 378, pp. 1–15, 2019, doi: 10.1016/j.cej.2019.122166.

M. Nurhayati, A. Linggawati, S. Anita, and T. A. Amri, "Preparation and characterization of calcium oxide heterogeneous catalyst derived from Anadara Granosa Shell for biodiesel synthesis," *ICoSE Conf. Instrumentation, Environ. Renew. Energy*, vol. 2016, pp. 1–8, 2016, doi: 10.18502/keg.v1i1.494.

M. S. Tizo, *et al.*, "Efficiency of calcium carbonate from eggshells as an adsorbent for cadmium removal in aqueous solution," *Sustain. Environ. Res.*, vol. 28, no. 6, pp. 326–332, 2018, doi: 10.1016/j.serj.2018.09.002.

M. S. Yafie and W. Widyastuti, "Sintering Terhadap Densitas Dan Kekerasan Pada MMC W - Cu Melalui Proses Metalurgi Serbuk," *J. Tek. POMITS*, vol. 3, no. 1, 2014, doi: 10.12962/j23373539.v3i1.5568.

M. W. Lukman, Y. Yudyanto, and H. Hartatiek, "Sintesis Biomaterial Komposit CaO-SiO_2 Berbasis Material Alam (Batuan Kapur dan Pasir Kuarsa) dengan Variasi Suhu Pemanasan dan Pengaruhnya Terhadap Porositas, Kekerasan dan Mikrostruktur," *J. Sains*, vol. 2, pp. 1–5, 2012.

M. Z. A. Khiri *et al.*, "The usability of ark clam shell (Anadara granosa) as calcium precursor to produce hydroxyapatite nanoparticle via wet chemical precipitate method in various sintering temperature," *Springerplus*, vol. 5, no. 1, 2016, doi: 10.1186/s40064-016-2824-y.

N. S. W. Supriyanto, Sukarni, P. Puspitasari, and A. A. Permanasari, "Synthesis and characterization of $CaO/CaCO_3$ from quail eggshell waste by solid state reaction process," *AIP Conf. Proc.*, vol. 2120, no. 040032, pp. 1–6. 2019, doi: 10.1063/1.5115670.

N. Tangboriboon, R. Kunanuruksapong, A. Sirivat, R. Kunanuruksapong, and A. Sirivat, "Preparation and properties of calcium oxide from eggshells via calcination,". *Mater. Sci. Pol.* vol. 30, pp. 313–322. 2012, doi: 10.2478/s13536-012-0055-7.

P. Kamalanathan, *et al.*, "Synthesis and sintering of hydroxyapatite derived from eggshells as a calcium precursor," *Ceram. Int.*, vol. 40, no. PB, pp. 16349–16359, 2014, doi: 10.1016/j.ceramint.2014.07.074.

P. M. Derlet, "Sintering Theory," *Powder Technology*. Villigen: Paul Scherer Institute, pp. 1–96. 2017.

P. Patnaik, *Handbook of Inorganic Chemicals*. New York: McGraw-Hill, 2001.

P. Puspitasari, *Sintesis Nanomaterial: Bottom Up dan Top Down*. Malang: Universitas Negeri Malang, 2017.

P. Puspitasari, H. F. N. Zhorifah, H. Suryanto, A. A. Permanasari, and R. W. Gayatri, "Synthesis and characterization of $CaCO_3/CaO$ from chicken eggshell with various calcination times," *Nanotechnol. Percept.*, vol. 15, pp. 65–71, 2019, doi: 10.4024/N22PU18A.ntp.15.01.

P. Puspitasari, V. Yuwanda, Sukarni, and J. W. Dika, "The properties of eggshell powders with the variation of sintering duration," *IOP Conf. Ser. Mater. Sci. Eng.*, vol. 515, no. 1, 2019, doi: 10.1088/1757-899X/515/1/012104.

R. Ahmad, R. Rohim, and N. Ibrahim, "Properties of waste eggshell as calcium oxide catalyst," *Appl. Mech. Mater.*, vol. 754–755, pp. 171–175, 2015, doi: 10.4028/www.scientific.net/amm.754-755.171.

R. Asthana, A. Kumar, and N. B. Dahotre, "Powder metallurgy and ceramic forming,". In: S. Joel, S. Grossman and M. Heidenry, Editors. *Materials Processing and Manufacturing Science*. Burlington, MA: Academic Press, 2006, pp. 167–245.

R. Bahanan, "Pengaruh waktu sonokimia terhadap ukuran kristal kalsium karbonat ($CaCO_3$)," *Thesis*. Universitas Islam Negeri Syarif Hidayatullah, Jakarta: 2010.

R. Mohadi, A. Lesbani, and Y. Susie, "Preparas Dan Karakterisasi Kaksium Oksida (CaO) Dari Tulang Ayam," *Chem. Prog.*, vol. 6, no. 2, pp. 76–80, 2013, doi: https://doi.org/1035799/cp.6.2.2013.3498.

S. B. Budihartomo, "Pengaruh Pressureleses Sintering Komposit AL-Kaolin Terhadapa Densitas, Kekerasan dan Struktur Mikro," *TRAKSI*, vol. 12, no. 1, pp. 1–14, 2012.

S. Gunasekaran, G. Anbalagan, and S. Pandi, "Raman and infrared spectra of carbonates of calcite structure," *J. Raman Spectrosc.*, vol. 37, no. 9, pp. 892–899, 2006, doi: 10.1002/jrs.1518.

S. L. Kang, *Sintering: Densification, Grain Growth and Microstructure*, 1st ed. Amsterdam: Elsevier, 2005.

S. Rujitanapanich, P. Kumpapan, and P. Wanjanoi, "Synthesis of hydroxyapatite from oyster shell via precipitation," *Energy Procedia*, vol. 56, no. C, pp. 112–117, 2014, doi: 10.1016/j.egypro.2014.07.138.

T. Habe, "Notes on the species of the genus Amusium (Mollusca)," *Bull. Natl. Sci.*, vol. 7, no. 1, pp. 1–5, 1964.

T. Pornchai, A. Imkum, and P. Apipong, "Effect of calcination time on physical and chemical properties of CaO – catalyst derived from industrial-eggshell wastes," *J Sci Technol MSU*, vol. 35, pp. 693–697, 2016.

T. W. Agustini, S. E. Ratnawati, B. A. Wibowo, and J. Hutabarat, "Pemanfaatan Ccangkang Kerang Simping (Amusium pleuronectes) Sebagai Sumber Kalsium Pada Produk Ekstrudat," *J. Pengolah. Has. Perikan. Indones.*, vol. 14, no. 2, pp. 134–142, 2011.

T. Witoon, "Characterization of calcium oxide derived from waste eggshell and its application as CO_2 sorbent," *Ceramics International.*, vol. 37, no. 2, pp. 3291–3298, 2011, doi: 10.1016/j.ceramint.2011.05.125.

W. Schatt and K.-P. Wieters, *Powder Metallurgy: Processing and Materials*. Belgium: EPMA, 1997.

Z. Xu, M. A. Hodgson, K. Chang, G. Chen, and X. Yuan, "Microstructure, weight loss and tensile properties of a powder metallurgical Fe-Mn-Si alloy," *MDPI*, vol. 7, no. 81, pp. 1–16, 2017, doi: 10.3390/met7030081.

3 Simulation for Oil Pan Production against Its Porosity, Shrinkage, and Niyama Criterion

Poppy Puspitasari,[1,2] Aulia Amar,[1] Suprayitno,[1]
Dewi 'Izzatus Tsamroh,[3] Samsul Ariffin Abdul
Karim,[4] and Mochamad Achyarsyah[5]*
[1] Faculty of Engineering, Universitas Negeri Malang, Indonesia
[2] Centre of Advanced Materials for Renewable Energy (CAMRY),
Universitas Negeri Malang, Indonesia
[3] Faculty of Engineering, University of Merdeka Malang,
Indonesia
[4] Software Engineering Programme, Faculty of Computing
and Informatics, Universiti Malaysia Sabah, Malaysia; Data
Technologies and Applications (DaTA) Research Group,
Universiti Malaysia Sabah, Malaysia
[5] Foundry Department, Bandung Polytechnic of Manufacturing,
Indonesia
*Corresponding author: Poppy Puspitasari, poppy@um.ac.id

CONTENTS

DOI: 10.1201/9781003320746-3

27

3.1   INTRODUCTION

Aluminum alloys over the past decade have been widely developed and used as an alternative of steel because of the lightweight properties (Kabnure, Shinde, & Patil, 2020). Aluminum-silicon alloy (Al-Si) is popularly used for casting machine components, electrical components, and automotive part because they have excellent thermal conductivity, prevent thermomechanical damage, and have a lower density, resulting in strong components (Goenka et al., 2019). Al-Si alloy is known as one of the excellent casting materials because of its superior casting characteristics that are attributed to its mechanical properties such as low density, high strength, good corrosion resistance, and good castability and because of its applicability to various types of castings (Patnaik, Saravanan, & Kumar, 2019).

This research requires relevant studies from previous researches. There are several studies that discuss the analysis and prediction of shrinkage and porosity defects in the casting process using the finite element method (FEM) simulation method. Previous research was published in research journals by Borlepwar and Biradar (2019); in this study the researchers used high-pressure die casting (HPDC) simulation using ProCAST software to identify shrinkage in the A380 aluminum oil pump casing casting process with variations of pin stroke, pin diameter, and casting pressure. The results showed that the optimal squeeze pin diameter was 9 mm, the optimal squeeze pin stroke was 17.5 mm, the optimal pressure to apply was 35 bar, and the optimal time for the squeeze pin actuation was 2.65 s.

Another research was carried out by Kwon and Kwon (2019), where HPDC method was simulated using the computer aided engineering (CAE) AnyCasting software in optimizing the manufacture of oil tanks in cars. Here SKD61 ($AlSi_2Cu_3$) material was used with initial casting temperature of 913.15 K. This study used variations in runner shape and shot sleeve speed, which aim to predict shrinkage defects in casting products. The results showed that the casting simulation software can identify the location of the casting defects so that they can be overcome by changing the location and design of the riser. Meanwhile, the shrinkage of the castings occurs in the hot-shrink area due to the nonuniform cooling temperature.

Subsequent research was carried out by Barot and Ayar (2019), where the identification of castings and solidification using casting simulations on aluminum alloy plates with variations in thickness was discussed. The results showed that the porosity values in experimental simulations had almost the same values and the casting yield values for plates with a thickness of 8, 12, and 16 mm were 91.20%, 91.5%, and 92.05%, while the casting quality was, respectively, 95.20%, 94.05%, and 93.68% and the depreciation that occurred was 10%, 5.88%, and 4.34%.

Subsequent research conducted by Qin, Su, Chen, and Liu (2019) determined the effect of HPDC casting parameters, which included molten metal temperature, low pressure speed, low-high pressure speed, and high pressure speed using the HPDC simulation method. The results showed that the high temperature of the molten metal and the low injection speed could reduce the pre-crystallization of the metal in the shot sleeve. But when the low pressure speed exceeds 0.2 m/s, there is an air trapped which cause porosity.

The HPDC casting method is one of the most widely used casting processes for producing automotive components with aluminum alloy materials (Kang, & Zhu,

2017). About 50% of global aluminum alloy casting products are manufactured using gravity die casting or HPDC processes (Abdulhadi, Aqida, & Ismail, 2019).

3.2   HIGH-PRESSURE DIE CASTING

HPDC is a metal casting method using a permanent mold where molten metal is injected into the mold cavity with high pressure (Jiao et al., 2019). The pressure used in the casting process is 7–350 MPa (1,015–50,763 lb/in^2). During the HPDC casting, the pressure will be maintained during the molten metal freezing process. The use of high pressure using hydraulic pistons in injecting molten metal into the mold cavity is the most important process in this casting process, which is what makes it different from other permanent mold casting processes. The HPDC casting process is basically a combination of a metal casting process and a forging process (due to pressure) (Groofer, 2019). HPDC casting is classified into two types based on the location of the furnace, namely, hot chamber and cold chamber (Cao et al., 2018).

3.2.1   HOT CHAMBER PROCESS

According to Kalpakjian, Schmid, and Sekar (2014), in casting HPDC hot chamber, metal is melted in a furnace attached to an HPDC machine, and a piston is used to inject molten metal under high pressure into the mold. The hot chamber process is usually used for materials with low melting points, such as zinc (Zn), Tin (Sn), and magnesium (Mg) alloys. The pressure used in hot chamber casting can reach up to 35 MPa, with an average pressure of 15 MPa.

The advantages of the hot chamber casting model are the speed in producing castings, reducing the turbulence that occurs, and minimizing the heat lost during the pressing process using a shot sleeve. The disadvantage of the hot chamber process is that it can be contaminated by other molten metals, and erosion of the shot sleeve can occur because it is submerged at high temperatures.

3.2.2   COLD CHAMBER PROCESS

In cold chamber die casting, the metal will be melted using a separate furnace from the die casting machine. The transfer of molten metal from the furnace to the die casting machine in this process can use a pump or ladle. The common cold chamber casting process utilizes pressures ranging from 20 to 140 MPa to inject the molten metal into the mold. The cold chamber method is commonly used for casting aluminum, brass, and magnesium alloys. The disadvantage of this process is that the production process is not as fast as the hot chamber process and turbulence often occurs (Groofer, 2019).

3.3   HIGH PRESSURE DIE CASTING PARAMETERS

The quality of the die casting product basically depends on the variation of process parameters, such as the metal casting temperature, slow/fast shot, and pressure; if this is not controlled properly, it can cause casting defects that can reduce casting quality. According to Ruzbarsky (2019), the main parameters in die casting that affect the characteristics of the casting results include pouring temperature, injection speed (slow/fast shot), and casting intensification pressure. The parameters that affect HPDC casting results are as follows.

1. Plunger Injection Speed

   The plunger injection speed ranks first among the main casting parameters that affect the quality of castings (Ruzbarsky, 2019). The injection speed of the plunger will affect the quality of the casting product. An inappropriate speed will cause uncontrolled turbulent flow and will cause the air to be trapped (Choukri, Nabil, & Said, 2019). The injection of molten metal in HPDC casting begins with inserting molten metal into the shot sleeve and then pushing using a plunger with two velocity phases, namely, slow shot velocity and fast shot velocity. The meaning of slow shot velocity and fast shot velocity is as follows.

   a. Slow shot velocity (first stage piston movement) is the initial speed; when metal is injected in this phase, the plunger is moved at a low speed until the molten metal fills the top of the shot sleeve and is then moved at a constant speed. This is intended to avoid turbulence in the molten metal due to the thrust that happened suddenly. The slow shot stage itself ends when the molten metal is in the area around the gate and will be followed by a fast shot.

   b. Fast shot velocity (second phase plunger movement) is the measurement of second velocity increment when molten metal is injected until it is completely filled (Dou, Lordan, Zhang, & Lane, 2019).

2. Pouring temperature

   Casting temperature is the important factors that must be considered in producing high-quality castings, because this factor greatly affects the quality of castings that include microstructure and mechanical properties, which results in castings having good physical properties. Pouring temperature is one of the variables in the casting process. This variable is important because if the pouring temperature is too low, then the mold cavity will not be completely filled where the molten metal will freeze first at the inlet. The metal casting temperature in HPDC casting for aluminum alloys varies between 650°C and 800°C. The fluidity of the molten metal will completely depend on the pouring temperature; the higher of pouring temperature, the fluidity is better and the lower the viscosity of the molten metal will result in filling the mold cavity rapidly (Akash & Harsimran, 2018).

3. Pressure intensification

   Pressure intensification is the final stage in the HPDC casting process which is carried out when the molten metal fills the mold cavity completely and is carried out during the metal solidification process in the mold. The amount of pressure during casting will affect the porosity value that occurs in the product. The higher the applied pressure, the lower the porosity. When pressure is injected, it will push out the gas in the liquid and the air in the gap between the mold wall and the molten metal. The volume of gas in the casting is inversely proportional to the pressure, where the higher the pressure, the smaller the gas volume, and the lower the porosity of the gas (Campbell, 2011). However, a too high pressure can result in the failure of the HPDC process, such as causing flash or liquid metal to come out of the mold, and can shorten the life of the HPDC machine.

The most common casting defects that occur in HPDC are porosity and shrinkage defect (Borlepwar & Biradar, 2019). Porosity is responsible for 70% of failures in the HPDC casting process of aluminum (Concer & Marcondes, 2017). Aluminum alloys can shrink between 3.5% and 6% during the solidification process, and porosity in aluminum caused by dissolved hydrogen gas or due to trapped air leads to forming cavities during solidification. HPDC casting quality is influenced by several factors, such as molten temperature, injection velocity, and pressure (Kwon & Kwon, 2019).

In current casting industries, the design and development of a casting layout is a trial and error method based on heuristic know-how. The solution achieved in this way lacks scientific calculation and analysis (Kwon & Kwon, 2019; El Mehtedi et al., 2020). Casting simulation methods can help to optimize designs, predict casting failures, and improve casting product yields (Kabnure et al., 2020). Casting simulation using the FEM is currently the most technically efficient and inexpensive technology for analyzing and evaluating metal casting processes (Kwon & Kwon, 2019; El Mehtedi et al., 2020).

3.4   CASTING SIMULATION

The use of simulation software has contributed significantly to the foundry industry and the development of casting products both in terms of quality improvement and economic value. The use of software is intended to reduce the risk of failure during the casting stage and to produce efficiency values in terms of manufacturing time and cost (Khan, 2018). The purpose of using this casting simulation software is to predict and identify the performance quality of a casting process. The comparison between the foundry design developments conventionally (trial and error) without casting simulation software has some tendency to repeat the process due to failure (Herbandono, 2011).

Currently there are several software developed for casting simulation, including MAGMASOFT, ProCAST, SolidCAST, NOVACAST, Altair Inspire Cast, ADSTEFAN, and so on. This study uses the Altair Inpire Cast metal casting software developed by Altair Inc. The origin is the computer-aided engineering software for casting simulations that can be applied to various gravity casting, tilt casting, investment casting, and high/low pressure die casting methods. Altair Inpire Cast has the ability to perform auto mesh generation from various file extensions and has a database management that can be adapted to the conditions of the casting to be performed, and therefore it can do the setting of casting parameters that are close to the actual conditions.

Altair Inpire Cast is able to perform casting simulations to predict casting defects, optimize the design of castings, and perform an integrated casting analysis. Based on this analysis, efforts can be made to prevent defects that may materialize and reduce the possibility of failure. Casting defects such as porosity and shrinkage can be predicted using solidification metrics such as the Niyama criterion using software simulations.

3.5   NIYAMA CRITERION

The identification of the Niyama criterion was first done by Dr. Niyama, a Japanese researcher, in 1982. The Niyama criterion is a compaction matrix in casting which is calculated at each mesh in the casting simulation to determine the location of porosity and shrinkage in the casting product. The Niyama criterion was originally developed to predict the location of porosity and shrinkage defects in steel materials and the results are very accurate, but currently the Niyama criterion is used for other materials, such as aluminum alloys and magnesium alloys (Ignaszak, 2017).

The main principle of the Niyama criterion is to compare the temperature gradient as a thermal parameter with the cooling rate which is very important in the metal freezing process, to predict the value of porosity and shrinkage using an interdendritic mechanism. The Niyama criterion equation can be written as follows:

$$N_y = \frac{G}{\sqrt{\dot{T}}} \qquad (3.1)$$

where
G = Thermal gradien (pouring temperature to solidification) (°C/cm)
$\dot{T}$ = Cooling rate (°C/s)

Niyama criterion (Figure 3.1) is a parameter that has no units (dimensionless). The parameters of the thermal gradient and cooling rate are reviewed at a certain temperature when freezing will be completed. This parameter will be easier to see during the simulation process using foundry simulation software. If in a casting simulation the Niyama criterion value is lower, it can be estimated that porosity will occur in that area. The Niyama criterion for each metal has a different value depending on the material and the type of alloy. Aluminum alloys have a critical point value of the Niyama criterion ranging from 0.25 to 0.3 (Monroe & Beckermann, 2014).

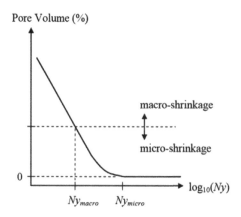

FIGURE 3.1   Niyama criterion (Ny) versus pore volume.

Several casting simulation software present Niyama criteria as output for analysis (Achyarsyah, Puspitasari, Budi, & Wicaksono, 2020). Niyama criterion is defined as the ratio of the thermal gradient to the square root of the cooling rate, both of which are evaluated at a specific temperature near the end of solidification (Equation 2.1). Niyama criterion is used to predict the porosity and shrinkage formed (Achyarsyah et al., 2020; Carlson & Beckermann, 2009; Carlson & Beckermann, 2009; Ignaszak, 2017).

3.6 MATERIAL AND METHODS

HPDC casting simulation process was done on Al-Si oil pan products using Altair Inspire Cast software. The design of the oil pan product was carried out using the Autodesk Inventor software with the geometry shown in Figure 3.2 and then imported into the Altair Inspire Cast software. The oil pan products are made using Al-Si$_{12}$ material with a casting mass of 0.53 kg (based on software calculations). Table 3.1 presents the properties of the Al-Si12 material used in this study obtained from the Altair Inspire Cast software. Simulation conditions were carried out by varying the pouring temperature and fast shot velocity (Table 3.2). This simulation process used a tetrahedral meshing which is divided using a self module from Altair Inspire cast. The simulation process carried out aims to obtain results from the porosity shrinkage, and

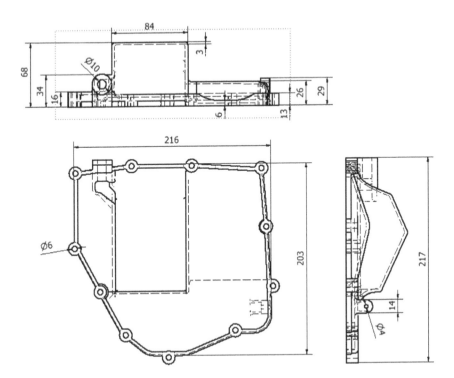

FIGURE 3.2 Design geometry of the oil pan.

TABLE 3.1
Data Material Al-Si$_{12}$

Properties	Value
Density	2680 kg/m^3 (20°C)
Thermal conductivity	130 W/m.K
Heat transfer coefficient	20
Specific heat	958 J/kg.K
Melting temperature	863 K
Dynamic viscosity	0.0018 kg/m.s

TABLE 3.2
Simulation Condition

Parameters	Value
Pouring temperature	700°C–800°C
Slow shot velocity	0.2 m/s
Fast shot velocity	2.5 m/s–3.5 m/s
Mold temperature (Steel H-13)	150 °C

the identification of the Niyama criterion. The results obtained are then tested for analysis of variance to determine the process parameters that contribute to the occurrence of porosity and shrinkage.

3.7 RESULTS AND DISCUSSION

3.7.1 POROSITY

Figure 3.3(a) shows the value of the pouring temperature on the porosity, where the porosity increases with the increasing of the pouring temperature. The lowest porosity value occurs at a pouring temperature of 700°C at 2.32%, while the highest porosity occurs at a pouring temperature of 800°C at 2.49%. This is due to the phenomenon of increasing the concentration of hydrogen gas dissolved in aluminum as the pouring temperature increases (Suprapto, 2011). According to Campbell, 2011), the concentration of hydrogen gas will increase as the temperature of aluminum increases, where the water content in the air and the water content in liquid aluminum will decompose and react to form elements of hydrogen (H_2) and oxygen (O), and then it will decompose into monatomic hydrogen (H) which will dissolve in the Al-Si alloy.

Figure 3.3(b) shows that there is a significant increase in the porosity value at a fast shot velocity of 3.0 m/s from 2.12% to 2.7%, which is due to an increase in velocity that will affect the turbulence of molten metal during the molten metal injection process at the shot slevee or mold cavity which results in air trapping which

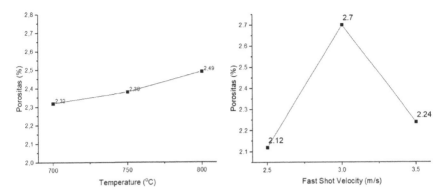

FIGURE 3.3   (a) Effect of pouring temperature on porosity. (b) Effect of fast shot velocity against porosity.

has the potential to increase porosity (Sharifi, Jamali, Sadayappan, & Wood, 2018; Tsoukalas, 2003). Furthermore, there was a decrease in the value of porosity at a fast shot velocity of 3.5 m/s to 2.24%, which was due to the high pressure and a decrease in the injection rate that occurred at the gate at high velocity (Vanli, 2010).

3.7.2   SHRINKAGE

Figure 3.4(a) shows the pouring temperature value against the shrinkage volume, where the highest shrinkage occurs at a pouring temperature of 700°C at 1411.50 mm³, then decreases at a pouring temperature of 750°C by 1254.38 mm³, and then at a pouring temperature of 800°C the volume shrinkage increased to 1308.28 mm³. These results indicate that the shrinkage is affected by the pouring temperature during the solidification process. High pouring temperature will lead to smaller shrinkage volume. This is because the higher the pouring temperature, the more irregular and evenly distributed the atomic structure in the aluminum, which will give a longer time in the perfect freezing process (Sujana & Setiawan, 2010). However, casting temperatures that are too high will form blowholes and shrinkage which reduce the properties and quality of the castings; the optimum temperature for casting aluminum alloys has a range between 700°C and 750°C (Ndaliman & Pius, 2007; Paul, 2014).

Figure 3.4 shows the shrinkage volume increase with the increasing of the fast shot velocity, where the lowest shrinkage volume occurs at a fast shot velocity of 2.5 m/s of 1317.43 mm³, while the highest value of volume shrinkage occurs at a fast shot velocity of 3.5 m/s of 1334.56 mm³. From Figure 3.4 it can be seen that the fast shot velocity affects the volume shrinkage value but not significantly. Although the highest value occurs at a fast shot velocity of 3.5 m/s of 1334.56 mm³, this is because an increase in the speed of HPDC casting will accelerate the molten metal filling the mold cavity and will result in a decrease in temperature which accelerates the premature solidification process. Fast shot velocity will affect the rate of premature solidification during the molten metal injection process into the mold (Outmani, Fouilland-Paille, Isselin, & El Mansori, 2017).

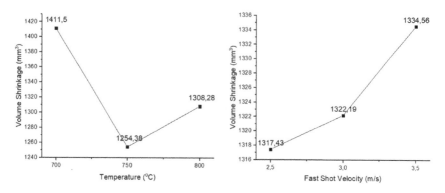

FIGURE 3.4   (a) Effect of pouring temperature on shrinkage. (b) Effect of fast shot velocity against shrinkage.

3.7.3 IDENTIFICATION OF NIYAMA CRITERION

Figure 3.5 shows the contours of the Niyama criterion which shows the location of the porosity that occurs in the HPDC casting of Al-Si oil pan products. Figure 3.5 shows the Niyama criterion with a pouring temperature of 700°C and a fast shot velocity of 2.5 m/s, having a minimum value of Niyama criterion as 0.0, which successfully predicts the location of the 2.07% porosity that is located in the center of the oil pan product. This also occurs in other parameter variations, where the minimum value of the Niyama criterion ranges from 0.00 to 0.01 and the location of the Niyama criterion is in accordance with the location of the porosity formed. The smaller the Niyama criterion, the greater the resulting porosity (Carlson & Beckermann, 2009; Ignaszak, 2017).

The poor prediction of the Niyama criterion occurred in the part of the oil pan product that had thicker dimensions (Figure 3.5). According to Kang et al. (2013), the Niyama criterion failed to predict shrinkage defects in thick areas due to the formation of a liquid pool during the solidification process (Kang, Gao, Wang, Ling, & Sun, 2013). Zhou et al. (2019) added that the dimensionless Niyama criterion was able to quantitatively predict low-porosity containing regions, but not in high-porosity containing regions (Zhou, Wu, Chen, & Han, 2019).

3.7.4 ANALYSIS OF VARIANCE

Table 3.3 shows the results of analysis of variance (ANOVA) based on each parameter (pouring temperature and *fast shot velocity*) against porosity. The pouring temperature contributed 8.65%. Meanwhile, fast shot velocity contributed 88.31% to porosity. This value means that the fast shot velocity variable has a significant effect on the formation of porosity in HPDC casting in the manufacture of Al-Si oil pan.

Table 3.4 shows the results of ANOVA based on each parameters (pouring temperature and *fast shot velocity*) against shrinkage volume. The pouring temperature has a contribution value of 96.20%. While the contribution value of the fast shot velocity parameter is 1.18% to the shrinkage formation. Based on this value, it can

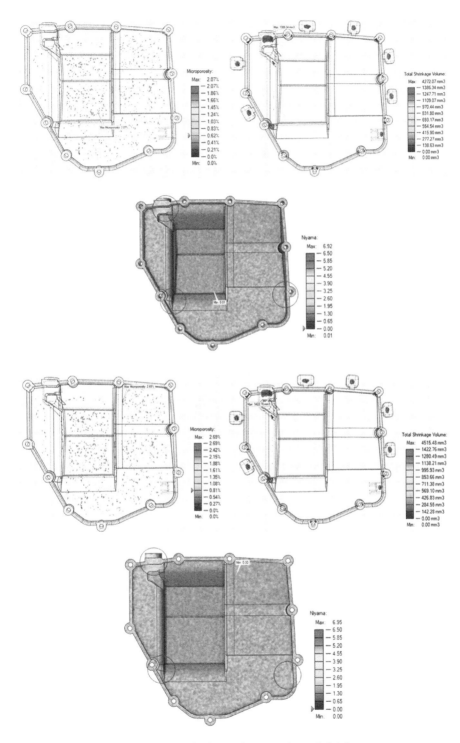

FIGURE 3.5 Correlation Niyama criterion against porosity and shrinkage.

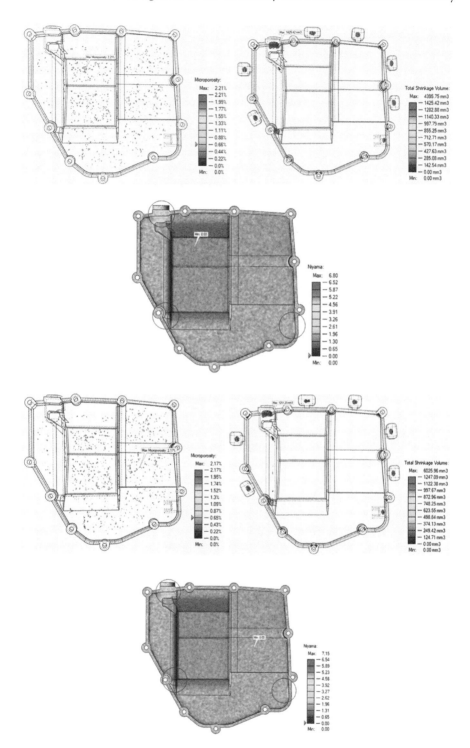

FIGURE 3.5   (Continued)

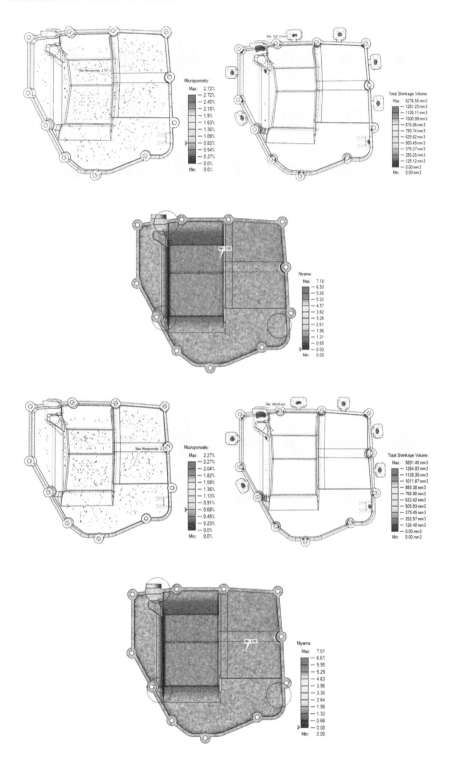

FIGURE 3.5 (Continued)

FIGURE 3.5 (Continued)

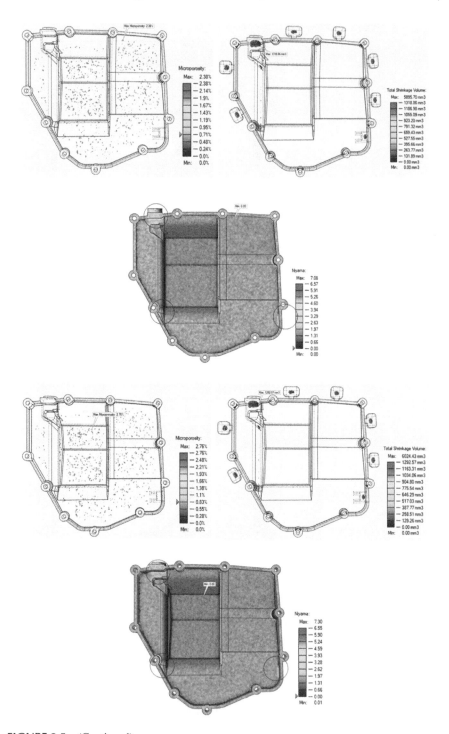

FIGURE 3.5   (Continued)

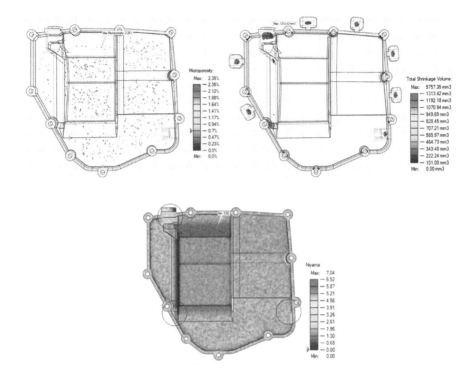

FIGURE 3.5 (Continued)

TABLE 3.3
Output ANOVA Porosity

Source	DF	Seq SS	Contribution	Adj SS	Adj MS	F-Value	P-Value
Pouring temperature	2	0.04616	8.65%	0.04616	0.023078	5.68	0.068
Fast shot velocity	2	0.47136	88.31%	0.47136	0.235678	58.03	0.001
Error	4	0.01624	3.04%	0.01624	0.004061		
Total	8	0.53376	100.00%				

TABLE 3.4
Output ANOVA Shrinkage

Source	DF	Seq SS	Contribution	Adj SS	Adj MS	F-Value	P-Value
Pouring temperature	2	38248.0	96.20%	38248.0	19124.0	73.41	0.001
Fast shot velocity	2	469.0	1.18%	469.0	234.5	0.90	0.476
Error	4	1042.0	2.62%	1042.0	260.5		
Total	8	39759.0	100.00%				

be interpreted that the pouring temperature parameter has a significant effect on the occurrence of shrinkage in HPDC casting in the manufacture of Al-Si oil pans.

3.8   CONCLUSION

- Based on research, the value of porosity increases with increasing pouring temperature. This is because the concentration of hydrogen gas dissolved in aluminum increases with increasing pouring temperature. There is an increase in porosity at a fast shot velocity of 3.0 m/s caused by trapped air due to turbulence, then porosity decreases at a fast shot velocity of 3.5 m/s due to high pressure and there is a decrease in the injection rate that occurs at the gate.
- ncreasing the pouring temperature decreases the volume shrinkage, because the high temperature will give a longer time in the perfect solidification process; however, a too high temperature will form a blowhole and shrinkage which reduces the nature and quality of the casting product. Increasing the fast shot velocity increases the shrinkage volume; this is because the metal will accelerate the molten metal filling the mold cavity and accelerate the premature solidification process.
- The Niyama criterion succeeded in predicting the location of the porosity in the oil pan. However, the Niyama criterion failed to predict shrinkage in thick sections or sections with high concentrations of porosity and shrinkage due to the formation of a liquid pool.

REFERENCES

Abdulhadi, H. A., Aqida, S. N., & Ismail, I. (2019). Tool failure in die casting. *Reference Module in Materials Science and Materials Engineering* (July). https://doi.org/10.1016/b978-0-12-803581-8.10483-7

Achyarsyah, M., Puspitasari, P., Budi, L. S., & Wicaksono, Y. A. (2020). Shrinkage simulation analysis on lifter cooler big at high temperature pouring liquid metal. *AIP Conference Proceedings*. https://doi.org/10.1063/5.0000985

Akash, T., & Harsimran, S. S. (2018). Application of die casting: A review paper. *Universal Review*, 7(3), 169–172.

Barot, R. P., & Ayar, V. S. (2019). Casting simulation and defect identification of geometry varied plates with experimental validation. *Materials Today: Proceedings*. https://doi.org/10.1016/j.matpr.2020.02.575

Borlepwar, P., & Biradar, S. (2019). Study on reduction in shrinkage defects in HPDC component by optimization of localized squeezing process. *International Journal of Metalcasting*, 13(4), 915–922. https://doi.org/10.1007/s40962-018-00295-9

Campbell, J. (2011). *Complete Casting Handbook: Metal Casting Processes, Techniques and Design*. Elsevier.

Cao, L., Liao, D., Sun, F., Chen, T., Teng, Z., & Tang, Y. (2018). Prediction of gas entrapment defects during zinc alloy high-pressure die casting based on gas-liquid multiphase flow model. *International Journal of Advanced Manufacturing Technology*, 94, 807–815.https://doi.org/10.1007/s00170-017-0926-5

Carlson, K. D., & Beckermann, C. (2009). Prediction of shrinkage pore volume fraction using a dimensionless Niyama criterion. *Metallurgical and Materials Transactions*

A: *Physical Metallurgy and Materials Science*, *40*(1), 163–175. https://doi.org/10.1007/s11661-008-9715-y

Choukri, K. M., Nabil, K. A. I., & Said, A. (2019). Effects of shot sleeve filling in HPDC machine. *Multidiscipline Modeling in Materials and Structures*, *15*(6), 1255–1273. https://doi.org/10.1108/MMMS-04-2019-0067

Concer, D., & Marcondes, P. V. P. (2017). Experimental and numerical simulation study of porosity on high-pressure aluminum die casting process. *Journal of the Brazilian Society of Mechanical Sciences and Engineering*, *39*, 3079–3088. https://doi.org/10.1007/s40430-016-0672-x

Dou, K. U. N., Lordan, E., Zhang, Y., & Lane, K. (2019). Influence of piston slow shot speed variation in shot sleeve on melt flow, solidification and defects formation in HPDC process: A numerical modelling study. *WSEAS Transaction on Heat and Mass Transfer*, *14*(2), 167–172.

El Mehtedi, M., Mancia, T., Buonadonna, P., Guzzini, L., Santini, E., & Forcellese, A. (2020). Design optimization of gate system on high pressure die casting of AlSi13Fe alloy by means of finite element simulations. *Procedia CIRP*, *88*, 509–514. https://doi.org/10.1016/j.procir.2020.05.088

Goenka, M., Nihal, C., Ramanathan, R., Gupta, P., Parashar, A., & Joel, J. (2019). Automobile parts casting-methods and materials used: a review. *Materials Today: Proceedings*, *22*, 2525–2531. https://doi.org/10.1016/j.matpr.2020.03.381

Groofer, M. (2019). Fundamentals of modern manufacturing. *Metallurgy of Welding*, 87–109. https://doi.org/10.1007/978-94-010-9506-8_6

Herbandono, K. (2011). *Universitas Indonesia Perancangan Dan Simulasi Pengecoran Pada Pembuatan Casing Turbin Uap Direct Condensing 3, 5 Mw Universitas Indonesia*.

Ignaszak, Z. (2017). Discussion on usability of the niyama criterion for porosity predicting in cast iron castings. *Archives of Foundry Engineering*, *17*(3), 196–204. https://doi.org/10.1515/afe-2017-0115

Jiao, X. Y., Wang, J., Liu, C., Guo, Z., Wang, J., Wang, Z., … Xiong, S. M. (2019). Influence of slow-shot speed on PSPs and porosity of AlSi 17 Cu 2.5 alloy during high pressure die casting. *Journal of Materials Processing Technology*, *268*, 63–69. https://doi.org/10.1016/j.jmatprotec.2019.01.008

Kabnure, B. B., Shinde, V. D., & Patil, D. C. (2020). Quality and yield improvement of ductile iron casting by simulation technique. *Materials Today: Proceedings*, *27*(xxxx), 111–116. https://doi.org/10.1016/j.matpr.2019.09.022

Kalpakjian, S., Schmid, S. R., & Sekar, K. S. V. (2014). *Manufacturing Engineering and Technology, Seventh Edition in SI Units*. Pearson Education South Asia Pte Ltd.

Kang, M., Gao, H., Wang, J., Ling, L., & Sun, B. (2013). Prediction of microporosity in complex thin-wall castings with the dimensionless niyama criterion. *Materials*, *6*(5), 1789–1802. https://doi.org/10.3390/ma6051789

Khan, M. A. (2018). A comparative study of simulation software. *International Journal of Simulation Model*, *17*, 197–209.

Kwon, H. J., & Kwon, H. K. (2019). Computer aided engineering (CAE) simulation for the design optimization of gate system on high pressure die casting (HPDC) process. *Robotics and Computer-Integrated Manufacturing*, *55*(January 2018), 147–153. https://doi.org/10.1016/j.rcim.2018.01.003

Monroe, C., & Beckermann, C. (2014). Prediction of hot tearing using a dimensionless Niyama criterion. *JOM*, *66*, 1439–1455. https://doi.org/10.1007/s11837-014-0999-7

Ndaliman, M., & Pius, A. (September 2007). Behavior of aluminum alloy castings under different pouring temperatures and speeds. *Leonardo Electronic Journal of Practices and Technologies*, *11*, 71–80.

Outmani, I., Fouilland-Paille, L., Isselin, J., & El Mansori, M. (2017). Effect of Si, Cu and processing parameters on Al-Si-Cu HPDC castings. *Journal of Materials Processing Technology*, *249*, 559–569. https://doi.org/10.1016/j.jmatprotec.2017.06.043

Patnaik, L., Saravanan, I., & Kumar, S. (2019). Die casting parameters and simulations for crankcase of automobile using MAGMAsoft. *Materials Today: Proceedings*, *22*(xxxx), 562–571. https://doi.org/10.1016/j.matpr.2019.08.208

Paul, Ager. I. A. (2014). Behavior of aluminum alloy castings under different pouring temperatures and speeds. *Discovery*, *22*(74), 62–71.

Qi, M. Fan, Kang, Y. Lin, & Zhu, G. Ming. (2017). Microstructure and properties of rheo-HPDC Al-8Si alloy prepared by air-cooled stirring rod process. *Transactions of Nonferrous Metals Society of China (English Edition)*, *27*(9), 1939–1946. https://doi.org/10.1016/S1003-6326(17)60218-8

Qin, X. Yu, Su, Y., Chen, J., & Liu, L. Jun. (2019). Finite element analysis for die casting parameters in high-pressure die casting process. *China Foundry*. https://doi.org/10.1007/s41230-019-8088-8

Ruzbarsky, J. (2019). *Al-Si Alloys Casts by Die Casting: A Case Study* (Vol. 1). Springer, pp. 1–71.

Sharifi, P., Jamali, J., Sadayappan, K., & Wood, J. T. (2018). Quantitative experimental study of defects induced by process parameters in the high-pressure die cast process. *Metallurgical and Materials Transactions A: Physical Metallurgy and Materials Science*, *49*(7), 3080–3090. https://doi.org/10.1007/s11661-018-4633-0

Sujana, W., & Setiawan, A. (2010). Pengaruh Temperatur Tuang Dan Waktu Tuang Terhadap Penyusutan Silinder Coran Alumunium Dengan Cetakan Logam. *Jurnal Flywheel*, *3*(1), 17–23.

Suprapto, W. (2011). *Universitas Indonesia Porositas gas..., Wahyono Suprapto, FT UI, 2011.*

Tsoukalas, V. D. (2003). The effect of die casting machine parameters on porosity of aluminium die castings. *International Journal of Cast Metals Research*, *15*(6), 581–588. https://doi.org/10.1080/13640461.2003.11819544

Vanli, A. S. (2010). Investigation the effects of process parameters on product quality in cold chamber high pressure die casting of magnesium alloys. *International Journal of Arts and Sciences*, *3*(9), 320–325.

Zhou, B., Wu, D., Chen, R., & Han, E. H. (2019). Prediction of shrinkage microporosity in gravity-cast and low-pressure sand-cast Mg–6Gd–3Y–0.5Zr magnesium alloys. *Advanced Engineering Materials*, *21*(12), 19–21. https://doi.org/10.1002/adem.201900755

4 Analysis of the Thermophysical Properties of SAE 5W-30 Lubricants with the Addition of Al$_2$O$_3$, TiO$_2$, and Hybrid Al$_2$O$_3$-TiO$_2$ Nanomaterials on the Performance of Motorcycles

Poppy Puspitasari,[1,2]* Syafira Bilqis Khoyroh,[1] Avita Ayu Permanasari,[1,2] Muhammad Mirza Abdillah Pratama[3], and Samsul Ariffin Abdul Karim[4]

[1] Faculty of Engineering, Universitas Negeri Malang, Indonesia
[2] Centre of Advanced Materials for Renewable Energy (CAMRY), Universitas Negeri Malang, Indonesia
[3] Civil Engineering Department, Universitas Negeri Malang, Indonesia
[4] Software Engineering Programme, Faculty of Computing and Informatics, Universiti Malaysia Sabah, Malaysia; Data Technologies and Applications (DaTA) Research Group, Universiti Malaysia Sabah, Malaysia
*Corresponding author: Poppy Puspitasari, poppy@um.ac.id

CONTENTS

DOI: 10.1201/9781003320746-4

4.1   INTRODUCTION

The efficiency and effectiveness of motor vehicle engine performance in the automotive industry is strongly influenced by the condition of the lubricating oil used. The ability of a lubricant to cope with changes in viscosity value to changes in temperature is called the viscosity index. A good lubricant is a lubricant that is not sensitive to changes in temperature starting from the engine starting until the engine performance increases. Lubricants have viscosity at high temperatures or low temperatures when the engine is operated, so the lubricant has its own grade (degree) regulated by the Society of Automotive Engineers (SAE) (Surbakti, 2019).

Nanoparticles have been suggested as lubricant additives to improve thermophysical performance. The addition of nanomaterials to the base fluid changes its thermophysical features, such as thermal conductivity, density, and viscosity (Mousavi & Zeinali Heris, 2020). In particular, research efforts have been directed toward using Titanium Dioxide (TiO_2) nanoparticles as additives that can be uniformly dispersed, and the results have shown that TiO_2 is able to reduce rolling forces and improve surface quality (Wu et al., 2019). Aluminum dioxide (Al_2O_3) is a good heat and electrical insulator. Al_2O_3 also plays an important role in the resistance of aluminum metal to rust with air.

Nanolubricants are nanoparticles in base lubricants that have a positive impact on tribological performance. The advent of nanotechnology has enabled the synthesis of nanomaterials in various fields (Azman, Syahrullail, & Rahim, 2018). Nanolubricant is useful for increasing the efficiency of the lubricant because it has a rolling effect so that it can reduce friction. In addition, nanoparticle additives will increase the absorption and heat transfer ability (Putro, 2007).

4.1.1   Previous Research

Several previous researches studied as theoretical studies carried out by several experts are described in Table 4.1.

4.1.2   Nanolubricant

Nanolubricant, which is a suspension of nanoparticles in a lubricant-based that has a positive impact on tribological performance. The advent of nanotechnology has

TABLE 4.1
Previous Research

Author	Results
Nugroho (2009)	Viscosity has a significant effect on temperature because the temperature of the lubricant increases and the viscosity decreases, and vice versa. In testing the thermal conductivity of new oil and used oil, the thermal conductivity of used oil is higher than that of the new oil. When testing motorcycle engine performance, the difference between new oil and used oil has little impact on torque, power, and fuel economy.
Suhanan, Kamal, Prayitno, Wiranata, and Pradecta (2016)	Based on the XRD test, the diffraction pattern showed that all 2θ angles of the peaks of TiO_2 nanoparticles had an anatase structure, which indicates that anatase titania is well dispersed in the base liquid and provides good stability. Moreover, the dynamic viscosity distribution is observed as an inverse function of temperature, and it is observed as a linear function of the volume fraction. However, the thermal conductivity decreased as the concentration of nanoparticles increased due to the sedimentation of the nanofluid during the experiment.
Johan (2009)	The results of characterization using XRD showed that the addition of TiO_2 did not affect the crystal structure of alumina and the reaction phase between TiO_2 and alumina was not detected. The results of density and hardness measurements showed that the addition of TiO_2 actually increased the density and hardness of the alumina material. The maximum density value is 3.9043 g.cm^{-3} obtained from samples that have been heated to a temperature of 1400°C then with the addition of TiO_2 up to 4%. The maximum hardness value is 1136.6 kg.mm^{-2}, obtained from the sample with the addition of 4% TiO_2 which has been heated to a temperature of 1500°C.
Septiadi et al. (2017)	The thermal conductivity character of the hybrid nanofluid Al_2O_3-TiO_2—Water at low and high concentrations showed an increase. The ratio that produces the highest thermal conductivity value is 75% Al_2O_3: 25% TiO_2 in both low and high concentrations. The highest increase from the low concentration was found in the volume fraction of 0.7%, namely, 32.50% and the highest increase from the high concentration was found in the volume fraction of 7%, namely, 41.07%. The best composition of nanoparticles with a significant increase in the thermal conductivity of hybrid nanofluids is at a low concentration with a volume fraction of 0.7%. This volume fraction (0.7%) with an increase in composition of six times can increase thermal conductivity of 20% and 32.5% to water, while the volume fraction of 7% requires a much larger increase in the composition, that is, 69 times.
Dinesh, Prasad, Kumar, Santharaj, and Santhip (2016)	This research is concerned with solving problems involving engine oil and its effective performance. MWCNTs and zinc oxide nanoparticles have volume fractions of 0.005% and 0.02%, respectively, dispersed with Gum Arabic surfactant in SAE 20W40 engine oil. Tribological properties such as friction coefficient and thermo-physical properties such as thermal conductivity, viscosity, flash point of the resulting nano-lubricant were evaluated and compared with the flash point of ordinary engine oil.
Yılmaz (2019)	In this study, copper (II) oxide (CuO), copper zinc iron oxide ($CuZnFe_2O_4$) and copper iron oxide ($CuFe_2O_4$) were used as nanoparticle additives in synthetic oil (5W-40) with a fraction of 0.08 wt%. Based on the research results, $CuZnFe_2O_4$ nano lubricants are the most prominent lubricants in terms of tribological performance, engine power, and CO emissions. There were an average 15% increase in engine power and an average 18% reduction in CO emissions.

enabled the synthesis of nanomaterials in various fields, including optics and optoelectronics, photocatalysis, electrical devices and sensors, biomedical applications, and so on due to their excellent chemical, thermal, and mechanical properties. Recently, significant research interest has been directed toward the synthesis of nanoparticles in base/lubricating fluids, as they can provide useful advantages in terms of heat transfer performance and tribological performance (Azman, Syahrullail & Rahim, 2018).

Lubricants are used to reduce friction and wear on moving engine components, such as engine pistons and gears. Another function is as a cooling medium and absorbs heat in moving components (Putro, 2007). Nanolubricant is a lubricant that is added with nanoparticle additives that have a rolling effect and can reduce the resulting friction; it is useful for increasing the efficiency of the lubricant. In addition, nanoparticle additives increase the ability to absorb and transfer heat and reduce the temperature of moving engine components (Putro, 2007).

4.1.3   LUBRICANT SAE 5W-30

Lubricants are products from petroleum refining that provide lubrication to bearings and rotating pistons in vehicle engines (Shoukat & Yoo, 2018). Lubricants have a viscosity level that has been determined by SAE. According to SAE, engine oils are classified as monograde or multigrade. Monograde engine oils are marked with a single number (20, 30, 40, 50, etc.). The figure shows the level of viscosity of the oil at a certain temperature. The grade numbering is directly proportional to the viscosity of the lubricant. The viscosity of engine oils designated by numbers only without the letter "W" (SAE 20, SAE 30, etc.) is determined at 212°F (100°C), while the viscosity of the engine lubricant followed by the letter "W" (SAE 20W, SAE 30W, etc.) indicates winter is determined at 0°F (−18°C) (Evans et al., 2013).

The viscosity of the engine oil is determined at high and low temperatures, which is called multigrade lubricant. They are designated by two numbers and the letter "W" (SAE 5W-30, SAE 15W-30, SAE 20W-50, etc.). The first number of the designation determines the viscosity of the lubricant at cold temperatures, and the second number determines the viscosity of the lubricant at high temperatures. For example, lubricant SAE 15W-30 has a low temperature viscosity similar to SAE 15, but has a high temperature viscosity similar to SAE 30. This provides a wide temperature range for multigrade lubricants (Evans et al., 2013). SAE 5W-30 is a premium quality engine oil formulated from synthetic base lubricants with selected additives and has the highest performance level in the API SN category, and contains lower impurities (Chowdhury et al., 2020). The properties of SAE 5W-30 are described in Table 4.2.

TABLE 4.2
Properties of Lubricant SAE 5W-30

Appearance	Density at 15°C (kg/m³)	Kinematic Viscosity at 40°C (cSt)	Viscosity Index	Flash Point (°C)	Pour point (°C)	Total Base No. mgKOH/g	Dynamic Viscosity (pa s)
Clear	851.5	63	171	226	−33	7.62	0.053

4.1.4 Thermophysical Properties

Thermophysical properties aim to determine the performance parameters, heat transfer coefficient, pressure drop, and thermal energy efficiency of a system. Thermal conductivity is considered to be the most important property of any fluid for heat transfer applications. It is directly related to the heat transfer coefficient which is related to system performance. Viscosity is a significant parameter for all fluid-related heat transfer applications (Nguyen et al., 2007). Viscosity is becoming an important transport phenomenon for the design of chemical processes. The performance of the heat exchanger is measured by the heat transfer coefficient, which is also affected by the viscosity as well as the distillation calculations and other heat transfer performances (Smith, Wilding, Oscarson, & Rowley, 2003). The stability of a suspension is related to the density of the particles. Density is required to calculate the required weight and space (volume). Specific heat is needed to analyze the energy and exergy performance of a system (Murshed, Leong, & Yang, 2016).

4.2 MATERIALS AND METHODS

The initial preparation stage is the process of characterizing Al_2O_3 and TiO_2 nanomaterials using XRD, scanning electron microscopy (SEM), and FTIR to determine the morphological structure, phase identification, and functional groups of the nanomaterials. Tables 4.3 and 4.4 show the specifications for Al_2O_3 and TiO_2 which are used as base lubricant additives.

The basic lubricant used is SAE 5W-30 lubricant. TiO_2 nanoparticles and 0.5 g Al_2O_3 nanoparticles were added to 900 ml of SAE 5W-30 lubricant as a base lubricant. Next, the stirring process was carried out using a magnetic stirrer for 4 hours at a speed of 1000 rpm. There are four samples of lubricants produced, namely, SAE

TABLE 4.3
Specifications of Aluminum Dioxide

Specification	Information
Brand	Sigma Aldrich
Size	< 10 nm
Purity	99.5%

TABLE 4.4
Titanium Dioxide Specification

Specification	Information
Brand	Sigma Aldrich
Size	< 25 nm
Purity	99.7%

5W-30 lubricant, nanolubricant SAE 5W-30 + TiO_2, nanolubricant SAE 5W-30 + Al_2O_3, and nanolubricant SAE 5W-30 + TiO_2 + Al_2O_3.

Testing of thermophysical properties includes density (ρ), dynamic viscosity (η), and thermal conductivity (k). The density of the nanolubricant can be determined by the mass divided by the volume. The mass is obtained from weighing the nanolubricant with a scale, while the volume is obtained using a measuring cup. Dynamic viscosity was obtained using the NDJ-8S viscometer. Thermal conductivity is one of the basic properties of materials obtained from the thermal conductivity analyzer—C – Therm / TCi. The nanolubricant was then tested for performance on motorcycle engine which included testing of engine torque and power.

4.3   RESULTS AND DISCUSSION

4.3.1   PHASE CHARACTERIZATION

XRD is a test that aims to analyze grain size, phase structure, and grain orientation. Phase identification of Al_2O_3 and TiO_2 is shown in Figure 4.1.

In addition to peak data, there are also crystal size values obtained using the Scherrer equation (Equation 4.1). The calculated values obtained are shown in Table 4.5 and Table 4.6.

$$D = \frac{k \times \lambda}{FWHM \times \cos \theta} \tag{4.1}$$

TiO_2 nanoparticles tested with test data range 10.01°–89.99°. The test results of TiO_2 nanoparticles showed compatibility with COD No. 96-500-0224, while the test results of Al_2O_3 nanoparticles showed a match with COD No. 96-100-0060.

4.3.2   MORPHOLOGICAL CHARACTERIZATION

SEM is a test method used to determine the morphology of a material. Figure 4.2(a) shows the results of SEM analysis of TiO_2 nanoparticles with a magnification of 100,000 times. Figure 4.2(b) is a surface morphology image of Al_2O_3 with a magnification of 1,000 times.

Measurements in Figure 4.2a get the grain average of 44.2 nm, which proves that the particle is a nanoparticle because its size is smaller than 100 nm. According to Komalasari and Sunendar (2013), TiO_2 nanoparticles are dispersed in a spherical form with a size of 100 nm and tend to agglomerate. The morphology of the TiO_2 nanoparticles is quite spherical so that it becomes a good rolling medium in the engine lubricant (Ali, Xianjun, Mai, et al., 2016).

The morphology of Al_2O_3 shows the phenomenon of aggregate and primary particle size. According to Teng, Hung, Jwo, Chen, & Jeng (2011), surface morphology of Al_2O_3 shows aggregate phenomena and primary particle size.

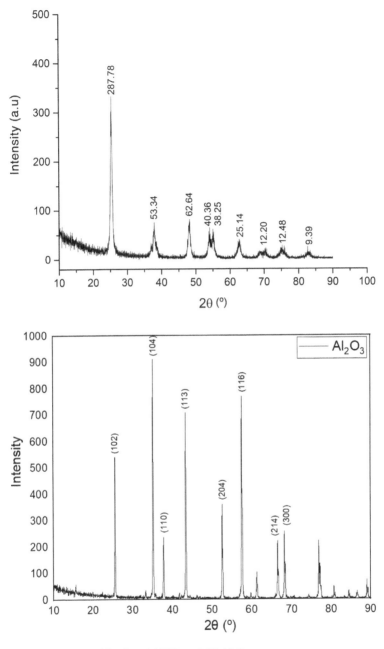

FIGURE 4.1   Phase identification: (a) TiO$_2$ and (b) Al$_2$O$_3$.

TABLE 4.5
Phase Identification from XRD TiO$_2$ Test Results

Position [°2Th]	FWHM [°2Th]	Crystallite Size [nm]
25.3520	0.3542	22.9891
37.8775	0.3936	21.3388
48.0901	0.5510	15.7879
53.9737	0.5510	16.1797
55.0680	0.4723	18.9689
62.7793	0.7085	13.1352
68.9990	0.9446	10.2049
70.3943	0.6293	15.4484
75.0743	1.1520	8.6967

TABLE 4.6
Phase Identification from XRD Al$_2$O$_3$ Test Results

Position [°2Th]	FWHM [°2Th]	Crystallite Size [nm]
25.5877	0.1180	69.0385
33.2528	0.1574	52.6742
35.1590	0.1378	60.4751
37.7873	0.1181	71.0982
41.7026	0.1181	71.9818
43.3549	0.0984	86.8789
46.2124	0.2362	36.5672
46.8451	0.3149	27.4935
52.5575	0.0720	123.0550
57.4924	0.1200	75.5081
57.6457	0.1200	75.5637
59.7768	0.3840	23.8619
61.3058	0.1440	64.1294
66.5148	0.1440	65.9742
66.6963	0.0960	99.0692
68.2008	0.1920	49.9682
68.3979	0.0960	100.0531
74.3334	0.3840	25.9615
76.8687	0.1440	70.4292
77.1336	0.1920	52.9191
80.6842	0.1440	72.3815
84.3450	0.1920	55.8292
86.4452	0.5760	18.9271
88.9986	0.1920	58.0105
89.2531	0.1440	77.5167

FIGURE 4.2 (a) TiO$_2$ with a magnification of 100,000x. (b) Al$_2$O$_3$ with a magnification of 1000x.

TABLE 4.7
TiO$_2$ Nanoparticle Functional Group

Band	Main Peak	Wave Number	Typical Band Assignment	Reference
1	3194.12	3196-3182	*Stretch vibration of the* O-H *bonds*	Asadi, Alarifi, and
2	3130/47	3133-3123		Foong (2020)
3	3072.6	3075-3064		
4	2308.79	2315-2304	O-H *stretching hydroxyl groups*	Devikala, Abisharani, and Rubavathi (2020)
5	1624.06	1637-1620	*Bending vibration of the* O-H *bonds*	Asadi et al. (2020)
6	1126.43	1151-1106	C-O *stretching vibration*	Asadi et al. (2020)
7	1045.42	1054-1041		
8	902.69	913-882	Si-O-H bonds	Wang, Zou, Li, Zou, and
9	810.1	817-806		Huang (2020)
10	754.17	759-747	*Stretching vibration of* Si-O-C	(Wang et al., 2020)
11	729.09	733-720		
12	667.37	669-661		

4.3.3 FOURIER-TRANSFORM INFRARED SPECTROSCOPY (FTIR) ANALYSIS

FTIR was used to identify elements and functional groups of TiO$_2$ and Al$_2$O$_3$ nanoparticles. The TiO$_2$ functional group is shown in Table 4.7, while the Al$_2$O$_3$ functional group is shown in Table 4.8.

4.3.4 THERMOPHYSICAL PROPERTIES

Density is an important thermophysical property; it is denoted by ρ (rho) and its unit is kg/m^3. Density is highly dependent on the nanoparticle materials used, while

TABLE 4.8

Al_2O_3 Nanoparticle Functional Group

Band	Main Peak	Wave Number	Typical Band Assignment	Reference
1	1276.88	1411-1263	*Carbon dioxide (OACAO) symmetric vibration*	Ates et al. (2015)
2	1246.02	1261-1236	*Triply degenerative v3 mode of SO_4^{-2} ion*	Manyasree, Kiranmayi, and Kumar (2018)
3	1058.92	1085-1047	*Bending vibration of the*	Sarmasti, Seyf, and
4	1043.49	1045-1022	*O-H bonds*	Khazaei (2021)
5	1018.41	1022-879	*Stretching vibrations of tetrahedrally coordinated Al-O*	Naayi, Hassan, and Salim (2018)
6	812.03	877-777	*Stretching vibration mode of the Al-O bond*	Naayi et al. (2018)
7	759.95	763-651	*Al-O bonds*	Naayi et al. (2018)
8	630.72	650-590	*Al-O stretching*	Manyasree et al. (2018)
9	547.78	588-534		
10	530.42	532-514	*Al-O stretching*	Manyasree et al. (2018)
11	486.06	503-445		

other parameters such as nanoparticle size, shape, zeta potential, and additives do not significantly affect nanofluid density. Solids have a greater density than liquids; therefore, the density of nanofluids increases with increasing concentration of nanoparticles. Like viscosity, density also decreases with increasing liquid temperature (Murshed et al., 2016).

$$\rho = \frac{m}{v} \tag{4.2}$$

where
ρ = Density of fluid (kg/m³)
m = Mass (kg)
v = Volume (m³)

Density of hybrid nanofluids can be measured using a density meter. In addition, some researchers measure density using volume and mass measurement techniques (Mercan, 2020). One example of measurement using a density meter is by Mahbubul et al. (2015), where nanofluid density is measured by the KEM-DA130N portable density meter (KYOTO, Japan). This device can measure density in the range of 0–2000 kg/m³, with an accuracy of ±1 kg/m³ and a resolution of 0.0001 g/cm³ (0.1 kg/m³). This tool can measure density in the temperature range of 0°C –40°C.

Figure 4.3 presents the density result that shows the highest value, namely, nanolubricant hybrid of 806 kg/m³. The density of nanolubricant is influenced by the density of the nanoparticles themselves. TiO_2 nanoparticles have a density of 4230 kg/m³ (Theivasanthi & Alagar, 2013), while the density of Al_2O_3 is 3980 kg/m³. This shows that the density value of the nanolubricant TiO_2 + SAE 5W-30 is greater than the

Density (kg/m³)

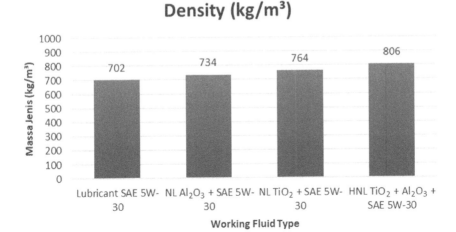

FIGURE 4.3   Density graph.

Dynamic Viscosity (Pa.s)

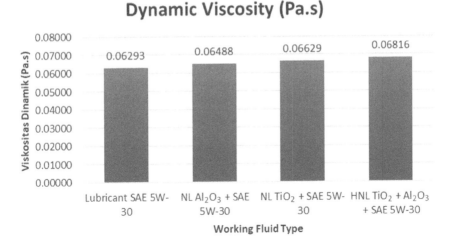

FIGURE 4.4   Dynamic viscosity graph.

nanolubricant Al_2O_3 + SAE 5W-30, and the highest density is the nanolubricant hybrid TiO_2 + Al_2O_3 + SAE 5W-30 (Tuan, Chen, Wang, Cheng, & Kuo, 2002).

Figure 4.4 shows a dynamic viscosity graph that has a slight difference from one sample to another, with the highest value being nanolubricant hybrid of 0.06816 Pa.s. according to Said et al. (2019). Viscosity value decreases with increasing temperature. The viscosity of the lubricant with the addition of nanoparticles will be higher than the base liquid, but the addition of TiO_2 turns out to have a higher viscosity value than the addition of Al_2O_3.

Figure 4.5 shows the thermal conductivity graph. The conductivity graph shows the highest values, namely, lubricant SAE 5W-30 of 0.130 W/mK, nanolubricant

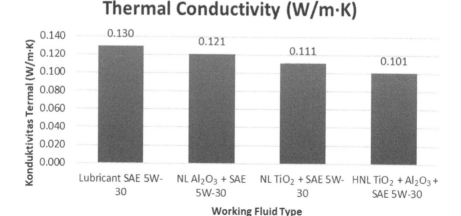

FIGURE 4.5 Thermal conductivity graph.

Al_2O_3 + SAE 5W-30 of 0.121 W/mK, nanolubricant TiO_2 + SAE 5W-30 of 0.111 W/mK, and nanolubricant hybrid TiO_2 + Al_2O_3 + SAE 5W-30 of 0.101 W/mK. According to Ahmed Ali, Xianjun, Abdelkareem, and Elsheikh (2019), thermal conductivity for nanolubricants and reference oils increases with increasing oil temperature. Factors affecting thermal conductivity are nanoparticle size, shape, type of base liquid, temperature and preparation technique will also affect thermal conductivity. The increase in thermal conductivity depends on the volume concentration and working temperature; therefore, the dispersion of these nanoparticles will significantly increase the thermal conductivity (Safiei, Rahman, Kulkarni, Ariffin, & Abd Malek, 2020).

4.3.5 Performance Test

The thermophysical properties which include density and dynamic viscosity also affect the torque value. This increase shows a more stable result against temperature changes and has a good effect on thinning resistance and film strength on the engine (Ali, Xianjun, Turkson, Peng, & Chen, 2016). Such a correlation or relationship can be interpreted that the higher the dynamic viscosity value, the higher the torque obtained, and the lower the dynamic viscosity value, the lower the torque obtained. This is because with a high dynamic viscosity value, the lubricant will be able to lubricate the deepest cracks in the combustion engine and is better able to reduce friction in the combustion chamber and produce a more optimal force (Antonius, Turnip, Atmadi, & Krisnamurti, 2019).

Figure 4.6 shows the result of the torque test. The graph shows the highest value, namely, nanolubricant hybrid of 8.24 Nm. Engine torque is a good indicator of the engine's ability to do work according to Ali, Xianjun, Mai, et al. (2016). With the addition of nanoparticles to the lubricant, it can effectively improve the tribological properties through the formation of a protective film on the surface and create a rolling effect between the friction surfaces. Al_2O_3 and TiO_2 are appropriate nanoparticles

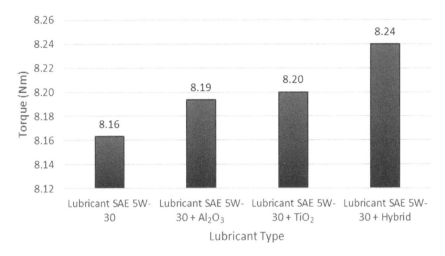

FIGURE 4.6   Graph of the effect of lubricant type on torque.

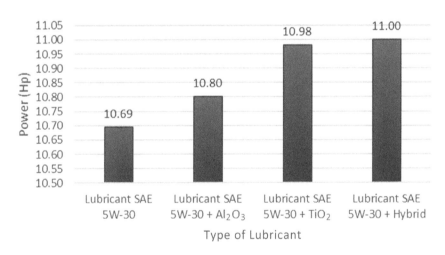

FIGURE 4.7   Graph of the effect of lubricant type on power.

due to their good tribological and thermal properties. Al_2O_3 nanoparticles are more effective in increasing anti-wear and abrasion resistance through the formation of a protective film, while TiO_2 nanoparticles are more effective in producing a rolling effect and reducing friction coefficient but more importantly stabilizing it (Ingole et al., 2013).

Figure 4.7 shows the result of the power test. From the four variations, the highest value for the nanolubricant hybrid was 11 Hp. The results show that the addition of Al_2O_3 and TiO_2 reduces friction losses compared to the base lubricant. Power losses are reduced because the nanoparticle mechanism converts shear into rolling friction. This reduction in friction loss can increase engine efficiency and power, and bearings between nanoparticles play a major role in reducing friction and wear which gives nanolubricant superior properties (Ahmed Ali et al., 2019). According to Ali et al., 2016 (Ahmed Ali et al., 2019), nanolubricant hybrid $TiO_2 + Al_2O_3$ can also accelerate the heating phase because it increases the sensitivity of the lubricant to temperature which results in reduced fuel consumption.

4.3.6 THE RELATIONSHIP OF NANOLUBRICANT THERMOPHYSICAL PROPERTIES WITH VEHICLE PERFORMANCE

Lubricants are used for cooling, sealing, and lubricating media. The additives themselves are used to maintain the overall performance of the lubricant, such as film formation, solidification, viscosity stabilizer, anti-corrosion, anti-wear, anti-friction, etc. (Kotia, Rajkhowa, Rao, & Ghosh, 2018). Based on the results of this study, it was shown that the nanolubricant hybrid $TiO_2 + Al_2O_3$ + SAE 5W-30 had an increasing impact on performance (torque and power). The increase in performance can be caused by several influencing factors. According to Kotia et al. (2018), the thermophysical properties of nanolubricant are affected by Brownian motion, particle agglomeration, agglomeration, interfacial layer, etc. Figure 4.8 presents the various mechanisms involved in augmentation in the thermophysical and tribological properties of the nanolubricant.

The heat transfer and flow behavior of the nanolubricant itself depends on its thermophysical properties. The viscosity of the nanolubricant will determine the shear force that occurs between adjacent fluid layers. The thermal conductivity of a lubricant plays a dominant role in determining its cooling behavior (Kotia et al., 2018). Most of the researchers have reported a linear increase in density with volume fraction of nanoparticles (Kotia & Ghosh, 2015). The specific heat in the lubricant contributes to the heat transfer performance of a nanolubricant (Elias et al., 2014). According to Kotia et al. (2018), dispersed nanoparticles will contribute to the repair, polishing, ball bearing, and third body mechanism, it will improve the tribological property improvement and improve the efficiency of automotive engines (Figure 4.9).

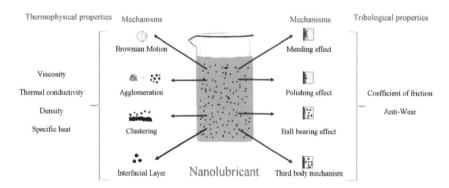

FIGURE 4.8 Graphical representation of the mechanisms involved with nanolubricant.

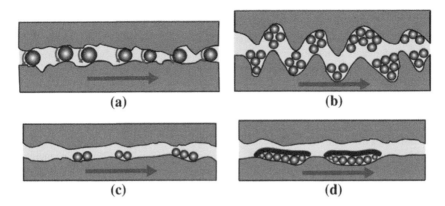

FIGURE 4.9 Illustration of (a) rolling mechanism, (b) self-repairing mechanism, (c) polishing mechanism, (d) tribo-film formation (Nurul Farhanah Azman & Samion, 2019).

Based on the results of the SEM test, it was shown that the morphology of the Al_2O_3 and TiO_2 nanoparticles was quite spherical. The spherical morphology of the nanoparticles will provide an effective rolling mechanism between the two sliding surfaces and change the friction mode from sliding to rolling, allowing the particles to reduce friction (Nurul Farhanah Azman & Samion, 2019). The addition of Al_2O_3 and TiO_2 nanoparticles to the SAE 5W-30 lubricant can effectively improve tribological properties which have a direct effect on increasing performance; reducing friction in the combustion chamber can produce more optimal force so as to increase engine efficiency and power.

4.4 CONCLUSION

1. The effect of adding TiO_2 and Al_2O_3 nanoparticles to the SAE 5W-30 lubricant on thermophysical properties has increased in the density of nanolubricant up to 806 kg/m³, an increase also occurred in the dynamic viscosity value up to 0.06816 Pa.s, and a decrease occurred in the thermal conductivity up to 0.101 W/mK. The data explains that the use of hybrid nanolubricant ($TiO_2 + Al_2O_3$) will increase the thermophysical properties of nanolubricant so that it will make a more optimal lubricant in its application to engine performance.
2. The addition of TiO_2 and Al_2O_3 nanoparticles to the SAE 5W-30 lubricant provides an increase in torque and power. Based on the results of the dyno test, there was an increase in torque using a nanolubricant hybrid ($TiO_2 + Al_2O_3$) up to 8.24 Nm, and an increase in power using a nanolubricant hybrid ($TiO_2 + Al_2O_3$) up to 11 HP.
3. The combination of nanolubricant hybrid ($TiO_2 + Al_2O_3$) produces the best performance (torque and power) for motorcycles. This can be seen from the increase in the results of the thermophysical properties and the results of the dyno test for hybrid nanolubricant ($TiO_2 + Al_2O_3$).

REFERENCES

Ahmed Ali, M. K., Xianjun, H., Abdelkareem, M. A. A., & Elsheikh, A. H. (2019). Role of nanolubricants formulated in improving vehicle engines performance. *IOP Conference Series: Materials Science and Engineering, 563*(2). https://doi.org/10.1088/1757-899X/563/2/022015

Ali, M. K. A., Xianjun, H., Mai, L., Bicheng, C., Turkson, R. F., & Qingping, C. (2016). Reducing frictional power losses and improving the scuffing resistance in automotive engines using hybrid nanomaterials as nano-lubricant additives. *Wear, 364–365*, 270–281. https://doi.org/10.1016/j.wear.2016.08.005

Ali, M. K. A., Xianjun, H., Turkson, R. F., Peng, Z., & Chen, X. (2016). Enhancing the thermophysical properties and tribological behaviour of engine oils using nano-lubricant additives. *RSC Advances, 6*(81), 77913–77924. https://doi.org/10.1039/c6ra10543b

Antonius, D., Turnip, K., Atmadi, P., & Krisnamurti, A. G. L. (2019). Analisis Pengaruh Jenis Pelumas Dasar Sintetik SAE 10W-40 Terhadap Daya, Torsi dan Konsumsi Bahan Bakar Mesin TIPE 2NR. *Jurnal METTEK, 5*(1), 10. https://doi.org/10.24843/mettek.2019.v05.i01.p02

Asadi, A., Alarifi, I. M., & Foong, L. K. (2020). An experimental study on characterization, stability and dynamic viscosity of CuO-TiO2/water hybrid nanofluid. *Journal of Molecular Liquids, 307*, 1–11. https://doi.org/10.1016/j.molliq.2020.112987

Ates, M., Demir, V., Arslan, Z., Daniels, J., Farah, I. O., & Bogatu, C. (2015). Evaluation of alpha and gamma aluminum oxide nanoparticle accumulation, toxicity, and depuration in Artemia salina larvae. *Environmental Toxicology, 30*(1), 109–118.

Azman, N. F., Syahrullail, S., & Rahim, E. A. (2018). Preparation and dispersion stability of graphite nanoparticles in palm oil. *Proceedinvs of Asia International Conference on Tribology, 19*(November), 132–141.

Azman, Nurul Farhanah, & Samion, S. (2019). Dispersion stability and lubrication mechanism of nanolubricants: a review. *International Journal of Precision Engineering and Manufacturing—Green Technology, 6*(2), 393–414. https://doi.org/10.1007/s40684-019-00080-x

Devikala, S., Abisharani, J. M., & Rubavathi, P. (2020). Preparation and characterization of TiO$_2$ nanoparticles using Cardiospermum halibacacabum leaves extract. *Materials Today: Proceedings, 40*, S189–S191. https://doi.org/10.1016/j.matpr.2020.08.582

Dinesh, R., Prasad, M. J. G., Kumar, R. R., Santharaj, N. J., & Santhip, J. (2016). Investigation of tribological and thermophysical properties of engine oil containing nano additives. *Materials Today Proceedings, 3*, 45–53. https://doi.org/10.1016/j.matpr.2016.01.120

Elias, M. M., Mahbubul, I. M., Saidur, R., Sohel, M. R., Shahrul, I. M., Khaleduzzaman, S. S., & Sadeghipour, S. (2014). Experimental investigation on the thermo-physical properties of Al$_2$O$_3$ nanoparticles suspended in car radiator coolant. *International Communications in Heat and Mass Transfer, 54*, 48–53. https://doi.org/10.1016/j.icheatmasstransfer.2014.03.005

Ingole, S., Charanpahari, A., Kakade, A., Umare, S. S., Bhatt, D. V., & Menghani, J. (2013). Tribological behavior of nano TiO$_2$ as an additive in base oil. *Wear, 301*(1–2), 776–785.

Johan, A. (2009). Karakterisasi Sifat Fisik dan Mekanik Bahan Refraktori _-Al$_2$O$_3$ Pengaruh Penambahan TiO$_2$. *Jurnal Penelitian Sains, 12*(2), 1–8.

Komalasari, M., & Sunendar, B. (2013). Penggunaan TiO$_2$ Partikel Nano Hasil Sintesis Berbasis Air Menggunakan Metoda Sol-Gel Pada Bahan Kapas Sebagai Aplikasi Untuk Tekstil Anti Uv. *Arena Tekstil, 28*(1), 16–21.

Kotia, A., & Ghosh, S. K. (2015). Experimental analysis for rheological properties of aluminium oxide (Al$_2$O$_3$)/gear oil (SAE EP-90) nanolubricant used in HEMM. *Industrial Lubrication and Tribology, 67*(6), 600–605.

Kotia, A., Rajkhowa, P., Rao, G. S., & Ghosh, S. K. (2018). Thermophysical and tribological properties of nanolubricants: a review. *Heat and Mass Transfer/Waerme- Und Stoffuebertragung, 54*(11), 3493–3508. https://doi.org/10.1007/s00231-018-2351-1

Mahbubul, I. M., Shahrul, I. M., Khaleduzzaman, S. S., Saidur, R., Amalina, M. A., & Turgut, A. (2015). Experimental investigation on effect of ultrasonication duration on colloidal dispersion and thermophysical properties of alumina-water nanofluid. *International Journal of Heat and Mass Transfer, 88*, 73–81. https://doi.org/10.1016/j.ijheatmasstransfer.2015.04.048

Manyasree, D., Kiranmayi, P., & Kumar, R. (2018). Synthesis, characterization and antibacterial activity of aluminium oxide nanoparticles. *Int. J. Pharm. Pharm. Sci., 10*(1), 32–35.

Mercan, H. (2020). Thermophysical and rheological properties of hybrid nanofluids. *Hybrid Nanofluids for Convection Heat Transfer*. Chapter 3, 101–142. https://doi.org/10.1016/b978-0-12-819280-1.00003-3

Mousavi, S. B., & Zeinali Heris, S. (2020). Experimental investigation of ZnO nanoparticles effects on thermophysical and tribological properties of diesel oil. *International Journal of Hydrogen Energy, 45*(43), 23603–23614. https://doi.org/10.1016/j.ijhydene.2020.05.259

Murshed, S. M. S., Leong, K. C., & Yang, C. (2016). Thermophysical properties of nanofluids. In *Handbook of Nanophysics: Nanoparticles and Quantum Dots*. Elsevier, 101–166. https://doi.org/10.1016/b978-0-12-821955-3.00003-0

Naayi, S. A., Hassan, A. I., & Salim, E. T. (2018). FTIR and X-ray diffraction analysis of Al_2O_3 nanostructured thin film prepared at low temperature using spray pyrolysis method. *International Journal of Nanoelectronics and Materials, 11*, 1–6.

Nguyen, C. T., Desgranges, F., Roy, G., Galanis, N., Maré, T., Boucher, S., & Angue Mintsa, H. (2007). Temperature and particle-size dependent viscosity data for water-based nanofluids—hysteresis phenomenon. *International Journal of Heat and Fluid Flow, 28*(6), 1492–1506. https://doi.org/10.1016/j.ijheatfluidflow.2007.02.004

Nugroho, M. A. (2009). *Analisa karakteristik viskositas dan konduktivitas termal oli mpx2 baru dan oli mpx2 bekas beserta pengaruhnya terhadap kinerja motor honda motor beat 110 cc tahun 2009*. 1–11.

Putro, D. A. (2007). *Analisis Sistem Pelumasan Pada Mesin Toyota Kijang Seri-5K*. Universitas Negeri Semarang.

Safiei, W., Rahman, M. M., Kulkarni, R., Ariffin, M. N., & Abd Malek, Z. A. (2020). Thermal conductivity and dynamic viscosity of nanofluids: a review. *Journal of Advanced Research in Fluid Mechanics and Thermal Sciences, 74*(2), 66–84.

Said, Z., Assad, M. E. H., Hachicha, A. A., Bellos, E., Abdelkareem, M. A., Alazaizeh, D. Z., & Yousef, B. A. A. (2019). Enhancing the performance of automotive radiators using nanofluids. *Renewable and Sustainable Energy Reviews, 112*, 183–194.

Sarmasti, N., Seyf, J., & Khazaei, A. (2021). Synthesis and characterization of [Fe_3O_4@CQDs@Si(CH_2)3NH_2@CC@EDA@SO_3H]+Cl- and Fe_3O_4@CQDs@Si(CH_2)3NH_2@CC@EDA@Cu nanocatalyts and their application in the synthesis of 5-amino-1,3-diphenyl-1H-pyrazole-4-carbonitrile and 1-(morpholino(phenyl)methyl)naphthalen-2-ol derivatives. *Arabian Journal of Chemistry, 14*(3). 103026. https://doi.org/10.1016/j.arabjc.2021.103026

Septiadi, W. N., Sukadana, I. G. K., Astawa, K., Nyoman, N. P. I. A., Trisdadewi, T., & Iswari, G. A. (2017). Konduktivitas Termal Hybrid Nanofluid Al 2 O 3 -TiO 2 -Air pada Fraksi Volume Rendah. *Prosiding Konferensi Nasional Engineering Perhotelan VIII - 2017, 2017*, 210–215.

Smith, G. J., Wilding, W. V., Oscarson, J. L., & Rowley, R. L. (2003). Correlation of liquid viscosity at the normal boiling point 1. *Young*. Paper presented at the Fifteenth Symposium on Thermophysical Properties, June 22-27, 2003, Boulder, Colorado, USA.

Suhanan, Kamal, S., Prayitno, Y. A. K., Wiranata, A., & Pradecta, M. R. (2016). Studi Eksperimental Sifat Termofisik Fluida Nano TiO2 /TermoXT-32. *National Symposium on Thermofluids VIII 2016, 32*(December), 1–6.

Surbakti, A. (2019). Pengaruh Jenis Oli Terhadap Konsumsi Bahan Bakar Pada Kendaraan Roda Dua 125 CC. *Piston, 4*(1), 8.

Teng, T.-P., Hung, Y.-H., Jwo, C.-S., Chen, C.-C., & Jeng, L.-Y. (2011). Pressure drop of TiO_2 nanofluid in circular pipes. *Particuology, 9*(5), 486–491.

Theivasanthi, T., & Alagar, M. (2013). Titanium dioxide (TiO_2) nanoparticles XRD analyses: An insight. *ArXiv Preprint ArXiv:1307.1091.*

Tuan, W. H., Chen, R. Z., Wang, T. C., Cheng, C. H., & Kuo, P. S. (2002). Mechanical properties of Al_2O_3/ZrO_2 composites. *Journal of the European Ceramic Society, 22*(16), 2827–2833.

Wang, Y., Zou, C., Li, W., Zou, Y., & Huang, H. (2020). Improving stability and thermal properties of TiO_2 nanofluids by supramolecular modification: High energy efficiency heat transfer medium for data center cooling system. *International Journal of Heat and Mass Transfer, 156,* 119735. https://doi.org/10.1016/j.ijheatmasstransfer.2020.119735

Wu, H., Jia, F., Zhao, J., Huang, S., Wang, L., & Jiao, S. (2019). Effect of water-based nanolubricant containing nano-TiO_2 on friction and wear behaviour of chrome steel at ambient and elevated temperatures. *Wear, 426–427*(November 2018), 792–804. https://doi.org/10.1016/j.wear.2018.11.023

Yılmaz, A. C. (2019). Effects of nano-lubricants on power and CO emission of a diesel engine: an experimental investigation. *Celal Bayar Üniversitesi Fen Bilimleri Dergisi, 15*(3), 251–256. https://doi.org/10.18466/cbayarfbe.453763

5 Heat Transfer Rate and Pressure Drop Characteristics on Shell and Tube Heat Exchanger with Graphene Oxide Nanofluid

Avita Ayu Permanasari,[1] Fajar Muktodi,*
Poppy Puspitasari,[3] Sukarni Sukarni,[4]
and Siti Nur Azella Zaine[5]
[1,2,3,4] Faculty of Engineering, Universitas Negeri Malang, Indonesia
[5] Department of Chemical Engineering, Universiti Teknologi PETRONAS, Malaysia
* Corresponding author: Avita Ayu Permanasari, avita.ayu. ft@um.ac.id

CONTENTS

DOI: 10.1201/9781003320746-5

5.1 INTRODUCTION

Nanofluid is a working fluid that has attracted the attention of researchers because it can improve thermal properties and flow characteristics. Based on research that has been conducted previously [1], the addition of nanoparticles to the base fluid can increase the thermal conductivity and conduction heat transfer coefficient. This advantage makes nanofluids to be used as a cooling fluid that can help improve the performance of heat exchangers [2]. Heat exchanger is an equipment used in various industries, such as the food industry, beverage industry, air conditioning, pharmaceutical industry, chemical industry, and power generation. Heat transfer characteristics are influenced by the convective heat transfer coefficient which is also influenced by the thermophysical properties of the fluid, such as density, viscosity, thermal conductivity, and specific heat [3]. The application of graphene oxide (GO) nanoparticles is a particle that can provide advantages due to its chemical structure and electrical properties. Research has shown that the use of TiO_2 nanofluids produces good heat transfer characteristics when compared to the use of Al_2O_3 nanofluids [3]. Albadr et al. studied the variation of the concentration of Al_2O_3 nanoparticles in water-based fluids in a shell and tube heat exchanger with horizontal type. The results showed an increase in heat transfer performance with the addition of the concentration of Al_2O_3 nanoparticles [4]. Ahmad et al. conducted research on the application of Multi Wall Carbon Nano Tube (MWCNT) nanoparticles to water-based fluids in shell and tube heat exchangers in a horizontal type compared to water [5]. The results showed an increase in the overall heat transfer coefficient using the MWCNT/water nanofluid [6].

    Fares et al. performed an analysis by application of GO nanofluid in shell and tube heat exchanger with variations in the concentration of GO nanoparticles of 0.01%, 0.05%, 0.1%, and 0.2%. The results showed that the addition of GO nanoparticles increased the thermal efficiency on the hot tube side by 24.4% and on the cold tube side by 7.3% [7]. In addition, research using GO nanoparticles has been carried out by Ghozatloo et al., with a nanoparticle concentration of 0.05%, 0.075%, and 0.1%. The results showed that the thermal conductivity of GO nanofluids increased by 15% (0.05% nanoparticle fraction), 29.2% (29.2% nanoparticle fraction), and 12.6% (0.1% nanoparticle fraction) at 25°C. Bhanvase et al. investigated the use of GO nanofluids in shell and tube heat exchangers with nanoparticle concentrations of 0.01% and 0.1%. The results showed that GO nanofluids with concentrations of 0.01% and 0.1% showed an increase in the thermal conductivity value of 9% and 20% higher than water at a temperature of 25°C [8].

    Milad et al. used exergy analysis to examine the effect of variations in the concentration of GO nanoparticles on thermal efficiency in shell and tube heat exchangers. The results showed an increase in heat transfer using GO nanofluid under laminar and turbulent flow conditions. Ghozatloo et al. conducted research on the application of GO/water nanofluids in a heat exchanger to the convective heat transfer coefficient in laminar flow. This study used GO nanoparticles with a concentration of 0.1%, where with that concentration there was an increase of 35.6% compared to the base fluid [7].

5.2 NANOFLUID

In recent years, nanofluids have become one of the most promising approaches to convert solar energy into steam energy through the absorption of nanoparticles. Compared with ordinary fluids, nanofluids exhibit better stability, much higher thermal conductivity, and higher heat transfer capabilities. Carbon-based nanofluids are widely used in the field of heat exchanger due to their high thermal conductivity and low cost [8].

Nanofluids are phase change materials (PCMs) of well-dispersed, stabilized, and highly conductive composites prepared by adding nanosized metal or nonmetallic particles to a base fluid, which can increase the heat transfer from the solution and reduce the cooling rate of the base fluid. Many studies on phase changes in nanofluids have been carried out, such as boiling and solidification. In general, nanofluids are accepted because they increase the heat transfer coefficient of nanoparticles.

Nanofluids are suspensions of nanoparticles with at least one dimension less than 100 nm, and therefore nanofluids are an excellent alternative for heat transfer fluids. Nanofluids, which are suspensions of nanoparticles in a fluid, have been found to have better thermal conductivity than individual base fluids. There are a number of advantages of using nanofluids over conventional fluids, such as long suspension stability, no clogging in the system, and less pressure drop. That is because the particle size of nanometer particles and the concentration of nanoparticles used are very small. In this study, nanofluids were used using GO/water. GO nanofluid is made by mixing GO flakes and distilled water with a certain concentration level of mass fraction. Figure 5.1 shows the process of making nanofluids using a two-step method [9, 10].

The first step is to prepare GO nanoflakes. Then the second step is mixing by dispersing the nanoflakes into water-aquadest with a mass fraction ratio of 0.04%, 0.05%, and 0.06% of the nanoparticles. The third step is that the nanofluids are stirred for approximately 30 minutes using a magnetic stirrer—this is done with

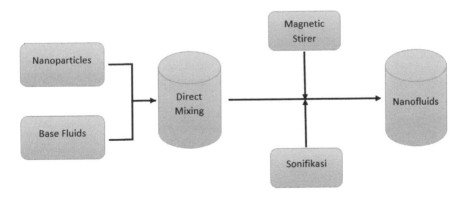

FIGURE 5.1   Synthesis of nanofluids using a two-step method.

the aim of facilitating the dispersion of nanoparticles into nanofluids. Then, after stirring with a magnetic stirrer, the nanofluid will be sonicated using a sonicator. In this sonification process, the time required is 1 hour with an amplitude of 100%. This process is carried out with the aim of minimizing the agglomeration of the nanoparticles. After the sonification process is complete, a cooling process is carried out on the nanofluid due to the sonification process making the nanofluid temperature to room temperature, which is 26°C. The cooling process aims to maintain the thermal conductivity value because the temperature in the nanofluid can affect the thermal conductivity [11].

5.3 GRAPHENE OXIDE

GO is a material that consists of a single two-dimensional carbon and a single sheet of sp2 carbon atoms combined. GO has good chemical properties, large surface area, and good optical, thermal, and electrical properties. Therefore, it is often applied to technological applications, such as solar cells and electrochemical energy storage devices. GO belongs to the carbon group. GO is a graphene nanosheet layer bonded to a graphene nanosheet bonded to an oxygen element. GO can be synthesized by graphite peeling or chemical oxidation methods. Graphene can be used as an adsorption material in catalytic applications because it has good adsorbent properties and photonic properties for catalysis because it can improve the photocatalytic properties of the material [12]. The specification of GO is shown in Table 5.1.

GO is in the category of graphite that can be processed by oxidation and mechanical methods. GO has begun to be widely used by researchers and industry to the global earth challenges and material demands. This happens because GO has thermophyscical properties that are interesting and different from other materials. GO presents a typical graphene-based carbon sheet structure followed by hydroxyl, carbonyl, carboxyl, and epoxy functional groups. In addition, GO can be dissolved in water. GO is a hybridized structure of sp3 and carbon. It is formed by a number of oxygen functional groups which make it hydrophilic by expansion between layers. GO has the ability to dissolve in organic materials, and this can also be done using other materials such as polymers and ceramics [14, 15].

TABLE 5.1
Graphene Oxide Material Specifications [13]

Graphene Oxide (GO)	Specifications
Thickness	0.8—1.2 nm
Purity	±99%
Bulk density	0.26 g/cm
Single layer ratio	±80%

Source: ACS Material, 2020.

5.4 HEAT EXCHANGER

This experimental study is equipped with a shell and tube heat exchanger. This heat exchanger used GO nanofluid as heating, which flows outside the pipe, while cold water flows in the inner pipe. The type of flow in the heat exchanger was counterflow. Figure 5.2 shows a schematic diagram of a shell and tube heat exchanger. Table 5.2 presents the specifications of the shell and tube heat exchanger used in this study.

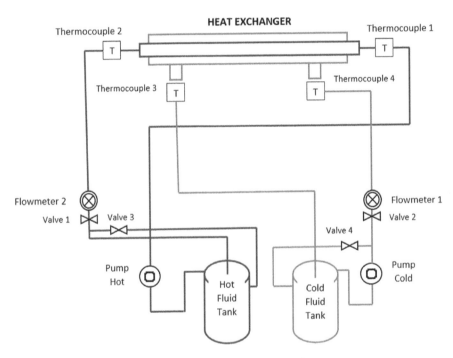

FIGURE 5.2 Schematic diagram of a shell and tube heat exchanger.

TABLE 5.2
Shell and Tube Heat Exchanger Specifications

Specifications	Dimension
Pipe Do = 21.45 mm P = 380 mm	1
Pipe Di = 8.42 mm P = 380 mm	4
Pump	2
Thermocouple DS18 B20 digital thermometer	2
Flowmeter	2
Heater	1
Flow rate cold water	1 l/min
Hot water flow rate	1 l/min

5.5 BASE FLUIDS

The base fluid that will be used to disperse the nanoparticles must be considered because if the base fluid is not correct, the value of high conductivity will not be achieved. The basic fluid that is often used in research is distilled water. When particles are added to the base fluid, the micrometer or nanometer size particles exhibit unique and enhanced properties when compared to the base fluid. These improved fluids can have significant applications in areas such as power generation, medicine, and in environmental processes. Table 5.3 presents the thermophysical properties of the base fluid used to disperse the GO nanoparticles.

5.6 THERMOPHYSICAL PROPERTIES

Thermophysical properties are nanofluids' physical properties as a medium in heat exchangers, which include density, specific heat, viscosity, and thermal conductivity. The density of nanofluids was measured using the Optima digital scale (OPD-E type) and a measuring cup to calculate the liquid's density. Specific heat and thermal conductivity were measured using the KD2 PRO with the SH-1 and KS-1 sensors. The SH-1 sensor consists of a 3-cm length and 1.3-mm diameter needle. In contrast, the KS-1 sensor consists of a needle with a length of 6 cm and a diameter of 1.3 mm. The needle was then inserted into the GO nanofluid until the process was detected by the KD2 PRO tool. The nanofluid's viscosity was measured using the NDJ-8S viscometer by inserting the rotor on the viscometer into the nanofluid, then determining the rotor according to the table in the manual book, after which the tool was turned on until the viscometer read the process.

5.6.1 HEAT TRANSFER CHARACTERISTICS

This study uses a shell and tube heat exchanger with a pump engine as the fluid mover. The type of convection used in this study was forced convection. The convective heat transfer coefficient was determined by calculating the Prandtl and Reynold numbers using Equation 5.1 and Equation 5.2 [16, 17].

$$Pr = \frac{\mu.Cp}{k} \qquad (5.1)$$

TABLE 5.3
Thermophysical Properties Base Fluid (Distilled Water)

Characters	Value	Measurement
Density	976,89	kg/m³
Specific heat	4178	J/kg. °C
Viscosity	0.00081	kg/m.s
Thermal Conductivity	0.547	W/m. °C

$$Re = \frac{v \cdot d \cdot \rho}{\mu} \qquad (5.2)$$

where Pr is the Prandtl number, C_p is the specific heat (J/kg°C), μ is the dynamic viscosity (kg/m.s), k is the thermal conductivity (W/m°C), Re is the Reynold number, v is the velocity of the fluid (m/s), d is the inner diameter of the pipe (m), and ρ is the density of the nanofluid (kg/m³). The convection coefficient is the propagation of fluid flow shown in the Nusselt number through Equation 5.3 and Equation 5.4 [16, 17].

$$Nu = 0,023\ Re^{0.8} Pr^{0.3} \qquad (5.3)$$

$$h = \frac{Nu.k}{D} \qquad (5.4)$$

where Nu is the Nusselt number, Re is the Reynold number, Pr is the Prandtl number, h is the convection coefficient (W/m²°C), D is the diameter (m), and k is the thermal conductivity (W/m°C).

The overall heat transfer coefficient was calculated using Equation 5.5 [16, 17].

$$U = \frac{1}{\dfrac{1}{hi} + \dfrac{\ln\left(\dfrac{Do}{Di}\right)}{2\pi k L} + \dfrac{1}{ho}} \qquad (5.5)$$

where U_a is the overall heat transfer coefficient (W/°C), D_o is the outer diameter of the pipe (m), D_i is the inner diameter of the pipe (m), h_i is the nanofluid heat transfer coefficient (W/m²°C), k is the thermal conductivity of the nanofluid (W/m°C), L is the length of the pipe (m). ΔT_{LMTD} was calculated using Equation 5.6 [16, 17].

$$\Delta T_{LMTD} = \frac{(Th,i - Tc,o) - (Th,o - Tc,i)}{\ln\left(\dfrac{Th,i - Tc,o}{Th,o - Tc,i}\right)} \qquad (5.6)$$

where $T_{c,i}$ and $T_{c,o}$ are, respectively, the temperature in which the cold stream enters and leaves, and $T_{h,i}$ and $T_{h,o}$ are, respectively, the temperature in which the heat flows in and flows out. The heat transfer of the nanofluids was calculated using Equation 5.7 [16, 17].

$$q = U \times A \times \Delta T_{LMTD} \qquad (5.7)$$

where A is the surface area of the heat exchanger (m²), q is the heat transfer rate (W), ΔT_{LMTD} is the difference in logarithmic mean temperature (°C). The pressure drop was calculated using Equation 5.8 and Equation 5.9 [18].

$$\Delta p = f \frac{L.\rho.v^2}{D.2} \tag{5.8}$$

where Δp is the pressure drop (Pa), f is the pipe friction factor, L is the pipe length (m), ρ is the density of the nanofluid (kg/m^3), v is the velocity of flow (m/s), and D is the inner diameter of the pipe (m). The friction factor was calculated using Equation 5.9.

$$f = \frac{64}{Re} \tag{5.9}$$

where f is the pipe friction factor and Re is the Reynold number [18].

5.7 RESULTS AND DISCUSSIONS

5.7.1 NANOFLAKES CHARACTERIZATION

The characterization of GO nanoflakes was carried out using the scanning electron microscope (SEM) and X-ray diffraction (XRD) tests. SEM-EDX testing is intended to determine the morphology of the form of GO nanoflakes at the micron scale along with the chemical elements contained in the GO nanoflakes. The SEM results of the GO nanoflakes are shown in Figure 5.3. The magnification used for the observation was 1000x. Figure 5.3 shows the measurement of GO nanoflakes at three different locations. Based on observations, at 3-grain points that have been identified, the grain size of GO nanoflakes has a size below 100 nm. The SEM-EDX test also analyzes the GO nanoflakes composition and the results of that analysis are shown in Table 5.4.

FIGURE 5.3 SEM test results for graphene oxide nanoflakes.

TABLE 5.4
Chemical Composition of Graphene Oxide Nanoflakes

Element	Wt (%)
C	56.173
O	37.01
Al	0.39
S	5.73
K	0.69

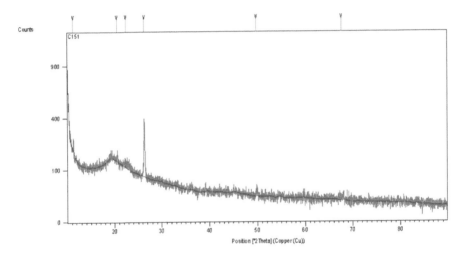

FIGURE 5.4   XRD test results for graphene oxide nanoflakes.

Figure 5.4 shows the XRD test, which aimed to identify the phase structure of GO nanoflakes. The highest diffraction peaks are shown at $2\theta = 10°$. This XRD structure follows the structural properties of GO reported by other researchers [19, 20].

5.7.2   THERMOPHYSICAL PROPERTIES

Measurement and calculation of the nanofluids' thermophysical properties that were applied to the heat exchanger include density, specific heat, viscosity, and thermal conductivity. Table 5.5 shows the results of measurements of the thermophysical properties of GO nanofluids. Figure 5.5(a) shows that the highest density value occurs at a concentration of 0.006% of 987.97 kg/m³ compared with variations in the concentration of GO nanoflakes of 0 0.004% and 0.005%. This happens due to the increasing mass of GO nanoflakes that is directly proportional to the growing value of nanofluid density. Figure 5.5(b) shows that the more the volume fraction of the nanoflakes is mixed in the basic fluid of distilled water, the lower the specific heating value. The

TABLE 5.5

Thermophysical Properties of Graphene Oxide Nanofluids

Nanofluids	Density (Kg/m³)	Specific Heat (J/kg°C)	Thermal Conductivity (W/Kg°C)	Viscosity (kg/m.s)
Water	976.8904	4178	0.547	0.00081
Mass fraction 0.004%	979.804	4167.73	0.563	0.00084
Mass fraction 0.005%	983.979	4143.83	0.583	0.00088
Mass fraction 0.006%	987.97	4131.36	0.593	0.0009

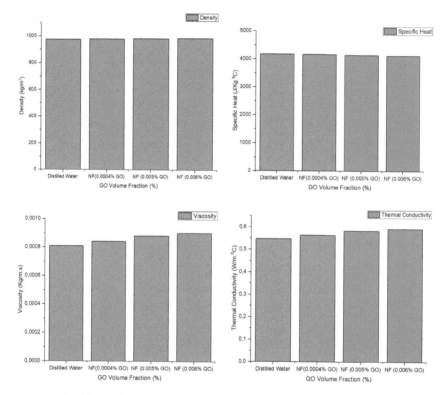

FIGURE 5.5 (a) Density with GO volume fraction variations, (b) specific heat with GO volume fraction variations; (c) viscosity with GO volume fraction variations; and (d) thermal conductivity with GO volume fraction variations.

most considerable specific heating value is in the working fluid without the addition of the concentration of GO nanoflakes of 4178 J/Kg.°C. Meanwhile, the smallest specific heat value is in the volume fraction of 0.006% of 4131.36 J/Kg.°C. Figure 5.5(c) shows that the viscosity increases 0.2% of the working fluid without the addition of GO nanoflakes toward a concentration of 0.006%, from 0.081 kg/m.s to 0.89 kg/m.s. According to the results, the more the concentration of GO nanoflakes increases, the denser the liquid properties of the nanofluid are. Figure 5.5(d) shows that the greatest thermal conductivity is found at the addition of 0.593 of 0.006% GO nanoflakes of W/m.°C and the lowest value is in the working fluid without the addition of nanoflakes of 0.557 W/m.°C [21].

5.7.3　Heat Transfer Characterization

5.7.3.1　Reynold Number and Nusselt Number

Table 5.6 shows the Reynold number and Nusselt number calculation with GO volume fraction and nanofluid flow rate variations. The addition of water flow rate at cold temperatures will increase the Reynold number. The Reynold number has the largest flow rate value of 0.9 l/min in the working fluid without GO nanoflakes. In the Reynold number in Figure 5.6(a), the value of each additional nanoflake

TABLE 5.6
Reynold and Nusselt Calculation

GO Volume Fraction	Nanofluid Flow Rate	Reynold Number	Nusselt Number
Distilled water	0.3	912.3239564	9.274423807
	0.6	1824.647913	16.14770974
	0.9	2736.971869	22.33489635
GO volume fraction 0.004%	0.3	882.3648073	9.043728557
	0.6	1764.729615	15.74604598
	0.9	2647.094422	21.77933036
GO volume fraction 0.005%	0.3	845.8462219	8.75842028
	0.6	1691.692444	15.24929542
	0.9	2537.538666	21.09224392
GO volume fraction 0.006%	0.3	830.4041367	8.636617703
	0.6	1660.808273	15.03722481
	0.9	2491.21241	20.79891595

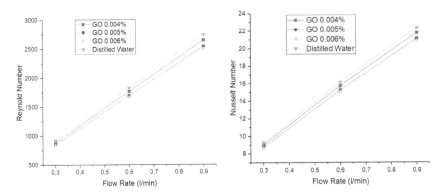

FIGURE 5.6　(a) Reynold number with nanofluid flow rate variations; (b) Nusselt number with nanofluid flow rate variations.

concentration decreases. This occurs because the more the mass of nanoflakes that are mixed in the working fluid, the smaller the value of the Reynold number. Meanwhile, Figure 5.6(b) shows that there is a decrease in the value of the Nusselt number as the volume fraction of GO nanoflakes increases, and the water flow rate increases on the cold side. The testing result shows that the flow rate and the addition of the volume fraction of GO nanoflakes to the Nusselt number which has the highest value, namely, the working fluid without the addition of GO nanoflakes with a water flow rate of 0.9 l/min which is equal to 22.3349. The lowest value of the Nusselt number is the working fluid with a 0.006% GO nanoflakes volume fraction and a flow rate of 0.3 l/min, namely 8.636618.

5.7.3.2   Convection Coefficient and Overall Heat Transfer Coefficient

Table 5.7 shows the convection coefficient and the overall heat transfer coefficient calculation with GO volume fraction and nanofluid flow rate variations. Figure 5.7(a) shows that there is a decrease in the value of the nanofluid convection coefficient due to the effect of flow rate and volume fraction. The results obtained from the convection coefficient suggest that the highest value is in the working fluid with the addition of 0.006% GO nanoflakes with a water flow rate on the cold side of 0.9 l/min of 1464.817 W/m².°C and the lowest value is in the working fluid, that is, the working fluid without the addition of GO nanoflakes is 602.5071 W/m².°C at a flow rate of 0.3 l/min. Meanwhile, Figure 5.7(b) shows that the overall heat transfer coefficient has the highest value for the working fluid with the addition of 0.006% GO nanoflakes at a flow rate of 0.9 l/min of 0.190533 W/°C, and the lowest value is in the working fluid without the addition of GO nanoflakes at a flow rate of 0.3 l/min of 0.187799 W/°C. It happens due to the influence of the Reynold and Nusselt numbers and the high thermal conductivity value of GO nanofluids with the addition of a volume

TABLE 5.7
Calculation of the Convection Coefficient and the Overall Heat Transfer Coefficient

GO Volume Fraction	Nanofluid Flow Rate	Convection Coefficient	Overall Heat Transfer Coefficient
Distilled water	0.3	602.5071048	0.187799241
	0.6	1049.025799	0.294664192
	0.9	1450.972483	0.3724497
GO volume fraction 0.004%	0.3	604.7053655	0.188390259
	0.6	1052.853193	0.295499621
	0.9	1456.266388	0.373414441
GO volume fraction 0.005%	0.3	606.432188	0.188854114
	0.6	1055.859766	0.296154934
	0.9	1460.424965	0.374170859
GO volume fraction 0.006%	0.3	608.2558548	0.190553416
	0.6	1059.034954	0.296846095
	0.9	1464.816765	0.374968339

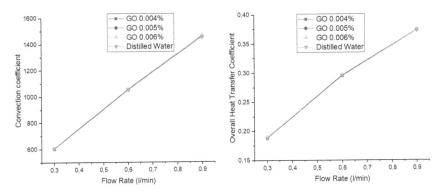

FIGURE 5.7 (a) Convection coefficient with nanofluid flow rate variations; (b) overall heat transfer coefficient with nanofluid flow rate variations.

concentration of 0.006% GO nanoflakes. According to Fares et al., increasing the concentration of graphene nanofluids increases the overall heat transfer coefficient significantly [7].

5.7.3.3 ΔT_{LMTD} and Heat Transfer

Figure 5.8(a) and Table 5.8 show that the lowest value of ΔT_{LMTD} is identified at the working fluid without the addition of GO nanoflakes from a flow rate variation of 0.3 l/min, namely 4.367500664, and the highest value of ΔT_{LMTD} is owned by nanofluids with a volume fraction of 0.006% from the variation of flow rate of 0.9 l/min, namely 6.698392799. This phenomenon occurs due to the influence of several factors, such as increased thermal conductivity and the addition of GO nanoflakes that are dispersed into the distilled water. Meanwhile, Figure 5.8(b) and Table 5.8 show that in heat transfer, the highest value is owned by the addition of 0.006% GO nanoflakes with a flow rate of 0.9 l/min worth 0.943062 watts and the lowest value is owned by the working fluid without the addition of GO nanoflakes with a flow rate. 0.3 l/min valued at 0.820213 watts. These results indicate that the working fluid with the addition of GO nanoflakes has a better heat transfer value than the working fluid without GO nanoflakes [22, 23].

5.7.3.4 Friction Factor and Pressure Drop

The calculation of friction factor and pressure drop shown at Table 5.9, while Figure 5.9(a) shows that the highest value of the friction factor is obtained at a variation of the concentration of GO nanoflakes 0.006% with a flow rate of 0.3 l/min of 0.077070907. Contrarily, the lowest is the distilled water working fluid without the addition of GO nanoflakes with a flow rate of 0.9 l/min 0.023383506. From the overall friction factor data obtained, it is clear that the value of the friction factor will decrease from each additional flow rate speed. The Reynold number's influence causes the phenomenon of decreasing the value of the friction factor because the greater the Reynold number's value, the smaller the value of the friction factor. Meanwhile, Figure 5.9(b) shows that the highest pressure drop value is obtained at the variation

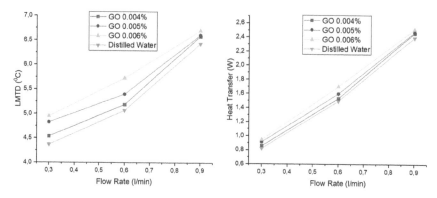

FIGURE 5.8   (a) ΔT_{LMTD} with nanofluid flow rate variations; (b) heat transfer with nanofluid flow rate variations.

TABLE 5.8
Calculation of ΔT_{LMTD} and Heat Transfer

GO Volume Fraction	Nanofluid Flow Rate	ΔT_{LMTD}	Heat Transfer
Distilled water	0.3	4.367500664	0.820213308
	0.6	5.068326188	1.493454242
	0.9	6.430404483	2.395002223
GO volume fraction 0.004%	0.3	4.538680652	0.855043225
	0.6	5.18694183	1.532739343
	0.9	6.587062761	2.459704361
GO volume fraction 0.005%	0.3	4.829155253	0.912005838
	0.6	5.398863121	1.598899954
	0.9	6.609366178	2.473032218
GO volume fraction 0.006%	0.3	4.949066887	0.9430616
	0.6	5.724617064	1.699330221
	0.9	6.698392799	2.511685225

of the concentration of GO nanoflakes 0.006% with a flow rate of 0.9 l/min, which is 41.6053635 Pa, whereas in the working fluid without the addition of GO nanoflakes with a flow rate of 0.3 l/min, the lowest pressure drop was 12.48160905 Pa. Pressure drop is related to viscosity, so the higher the nanofluid viscosity value, the higher the pressure drop value in the nanofluid [24].

5.8   CONCLUSIONS

Optimization of the heat exchanger using GO/water nanofluid as a working fluid shows promising results to improve the heat exchanger's performance. It can be concluded as follows: The results of the SEM-EDX test show the amount of chemical

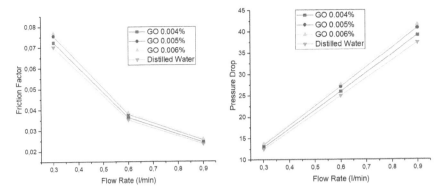

FIGURE 5.9   (a) Friction factor with nanofluid flow rate variations; (b) pressure drop with nanofluid flow rate variations.

TABLE 5.9
Calculation of Friction Factor and Pressure Drop

GO Volume Fraction	Nanofluid Flow Rate	Friction Factor	Pressure Drop
Distilled water	0.3	0.07015052	12.48160905
	0.6	0.03507526	24.9632181
	0.9	0.023383507	37.44482715
GO volume fraction 0.004%	0.3	0.072532358	12.94389087
	0.6	0.036266179	25.99809103
	0.9	0.024177453	39.15530818
GO volume fraction 0.005%	0.3	0.075663872	13.50273073
	0.6	0.037831936	27.12053324
	0.9	0.025221291	40.84580041
GO volume fraction 0.006%	0.3	0.077070907	13.75382571
	0.6	0.038535454	27.62486308
	0.9	0.025690302	41.6053635

element content. The XRD pattern graph has the highest peak at $2\theta = 10.14°$. This XRD structure is in accordance with the structural properties of GO reported by other researchers. Heat transfer characteristics at Reynold and Nusselt numbers have values that frequently decrease with the increasing volume fraction of GO nanoflakes. The addition of variations in the volume fraction of GO nanoflakes in the basic fluid of distilled water affects the increase in GO nanofluids' heat transfer rate. It obtains the highest value in the variation of the concentration of GO nanoflakes 0.006% with a flow rate of 0.9 l/min, which is 2.511685 W. The addition of variations in the volume fraction of GO nanoflakes on the base fluid of distilled water affects the increase in the value of pressure drop (Δp). The highest pressure drop value occurred at the variation of the concentration of GO nanoflakes 0.006%, namely 41.14308168 Pa.

ACKNOWLEDGMENT

This research is supported by PNBP Universitas Negeri Malang 2020 with Contract No. 3.3.16/UN32/KP/2020.

REFERENCES

[1] M. Vakili, S. Khosrojerdi, P. Aghajannezhad, and M. Yahyaei, A hybrid artificial neural network-genetic algorithm modeling approach for viscosity estimation of graphene nanoplatelets nanofluid using experimental data (*Int. Commun. Heat Mass Transf*). **82**, 40 (2017).

[2] R. Taherialekouhi, S. Rasouli, and A. Khosravi, An experimental study on stability and thermal conductivity of water-graphene oxide/aluminum oxide nanoparticles as a cooling hybrid nanofluid (*Int. J. Heat Mass Transf*). **145**, 118751 (2019).

[3] F. Sedaghat and F. Youse, Synthesizes, characterization, measurements and modeling thermal conductivity and viscosity of graphene quantum dots nanofluids (*J. Mol. Liq.*) **278**, 299 (2019).

[4] K. Zlaoui, N. Sdiri, D. Jellouli, and K. Horchani-naifer, Investigations on electrical conductivity and dielectric properties of graphene oxide nanosheets synthetized from modified Hummer's method (*J. Mol. Struct.*) **1216**, 128304 (2020).

[5] H. Li, H. Huang, G. Xu, J. Wen, and H. Wu, Performance analysis of a novel compact air-air heat exchanger for aircraft gas turbine engine using LMTD method (*Appl. Therm. Eng.*) **116**, 445 (2017).

[6] A. Hajatzadeh, S. Aghakhani, M. Afrand, and B. Mahmoudi, An updated review on application of nanofluids in heat exchangers for saving energy (*Energy Convers. Manag.*) **198**, 111886 (2019).

[7] M. Fares, M. Al-mayyahi, and M. Al-saad, Heat transfer analysis of a shell and tube heat exchanger operated with graphene nanofluids (*Case Stud. Therm. Eng.*) **18**, 100584 (2020).

[8] A. Ghozatloo, A. Rashidi, and M. Shariaty-niassar, Convective heat transfer enhancement of graphene nanofluids in shell and tube heat exchanger (*Exp. Therm. Fluid Sci.*) **53**, 136 (2013).

[9] B.A. Bhanvase, D.P. Barai, S.H. Sonawane, N. Kumar, and S.S. Sonawane, *Intensified Heat Transfer Rate With the Use of Nanofluids* (*Handbook of Nanomaterials for Industrial Applications*, Elsevier Inc., 2018).

[10] A.A. Permanasari, B.S. Kuncara, and P. Puspitasari, Convective heat transfer characteristics of TiO2-EG nanofluid as coolant fluid in heat exchanger (*AIP Conf. Proc.*) **2120**, 05001 (2019).

[11] D.A. Firlianda, A.A. Permanasari, and P. Puspitasari, Heat transfer enhancement using nanofluids (MnFe2O4- ethylene glycol) in mini heat exchanger shell and tube (*AIP Conf. Proc.*) **2120**, **05001** (2019).

[12] E. Burresi, N. Taurisano, and M.L. Protopapa, Influence of the synthesis conditions on the microstructural, compositional and morphological properties of graphene oxide sheets (*Ceram. Int.*) **46**, 22067 (2020).

[13] O. Hazards, Safety data sheet – Single layer graphene (*ACS Mater. LLC.*) **1** (2017).

[14] R. Ikram, B. Mohamed, and W. Ahmad, An overview of industrial scalable production of graphene oxide and analytical approaches for synthesis and characterization (*Integr. Med. Res.*) **9**, 11587 (2020).

[15] M.D.A. Pereira-da-silva and F.A. Ferri, *Scanning Electron Microscopy* (Elsevier Inc., 2017).

[16] Z. Hong, J. Pei, Y. Wang, B. Cao, M. Mao, and H. Liu, Characteristics of the direct absorption solar collectors based on reduced graphene oxide nanofluids in solar steam evaporation *(Energy Convers. Manag.)* **199,** 112019 (2019).

[17] N. Lertcumfu, P. Jaita, S. Thammarong, S. Lamkhao, S. Tandorn, C. Randorn, T. Tunkasiri, and G. Rujijanagul, Influence of graphene oxide additive on physical, microstructure, adsorption, and photocatalytic properties of calcined kaolinite-based geopolymer ceramic composites *(Colloids Surfaces A Physicochem. Eng. Asp.)* **602,** 125080 (n.d.).

[18] K. Moses, R.K. Koech, O. Kingsley, I. Ojeaga, K. Gabriel, and P.A. Onwualu, Synthesis and characterization of graphene oxide from locally mined graphite flakes and its supercapacitor applications *(Results Mater.)* **7,** 100113 (2020).

[19] S. Korkmaz and İ. Af, Graphene and graphene oxide based aerogels: Synthesis, characteristics and supercapacitor applications *(Elsevier.com)* **27** (2020).

[20] S. Iranmanesh, M. Mehrali, E. Sadeghinezhad, B.C. Ang and H.C. Ong, Evaluation of viscosity and thermal conductivity of graphene nanoplatelets nanofluids through a combined experimental–statistical approach using respond surface methodology method *(Int. Commun. Heat Mass Transf.)* **79,** 74 (2016).

[21] F. Ling , L. Yixin , M. Pengjv and G. Yongkun, Supercooling and heterogeneous nucleation in acoustically levitated deionized water and graphene oxide nanofluids droplets *(Exp. Therm. Fluid Sci.)* **103,** 143 (2019).

[22] J. Epp, *X-Ray Diffraction (XRD) Techniques for Materials Characterization* (Elsevier Ltd, 2016).

[23] M. Rabbani and E. Mohseni, Exergy analysis of a shell-and-tube heat exchanger using graphene oxide nanofluids *(Exp. Therm. Fluid Sci.)* **83,** 100 (2017).

[24] A.O. Morais and J.A.W. Gut, Determination of the effective radial thermal diffusivity for evaluating enhanced heat transfer in tubes under non-Newtonian laminar flow *(Braz. J. Chem. Eng.)* **32,** 445 (2015).

6 Microstructure Change of Aluminum 6061 through Natural and Artificial Aging

Dewi 'Izzatus Tsamroh,[1] Poppy Puspitasari,[2,3]*
Muchammad Riza Fauzy,[4] Agus Suprapto,[5]
and Pungky Eka Setyawan[6]
[1,5,6] Faculty of Engineering, University of Merdeka Malang, Indonesia
[2] Faculty of Engineering, Universitas Negeri Malang, Indonesia
[3] Centre of Advanced Materials for Renewable Energy (CAMRY), Universitas Negeri Malang, Indonesia
[4] Faculty of Engineering, University of Merdeka Malang, Indonesia
* Corresponding author: Dewi 'Izzatus Tsamroh, izza@unmer.ac.id

CONTENTS

DOI: 10.1201/9781003320746-6

81

6.1 INTRODUCTION: BACKGROUND

Aluminum alloy is a kind of material that is widely used in the world. Recently, the use of aluminum has high demand and it will increase rapidly over the years (Puspitasari et al., 2016). This can be up to 9.9% per year in tons (Tsamroh, 2021). In Indonesia, the Ministry of Industry targeted that Indonesia should be able to produce up to 1.5–2 million tons of aluminum by 2025 (Indonesia, 2018). This high demand is due to the ongoing development of the manufacturing industry. The high use of aluminum is attributed to its several beneficial properties, including being lightweight, ductile, and resistant to corrosion; besides, it can also be recycled (Woodford, 2021). Figure 6.1 shows the various uses of aluminum.

From Figure 6.1, it can be known that aluminum is widely used as a transportation component. For example, the aerospace industry uses aluminum for up to 90% of its components (Rambabu et al., 2017). Many kinds of automotive components are made of aluminum, such as valves, engine blocks, etc. (Ogunsemi et al., 2021). Due to its widespread use in various fields, aluminum needs to be improved in its properties continuously. The increasing use of aluminum alloys certainly affects the production and consumption of aluminum in the world. Recently, research on aluminum alloy has been conducted in many countries, for example, China, the USA, UAE, India, etc. (Woodford, 2021).

Aluminum has been classified into seven series. One of them is aluminum 6061, which is a part of series 6xxx. Aluminum 6061 consists of aluminum-magnesium-silicon, one of the treatable heat alloys, and has medium strength (Rajasekaran et al., 2012). However, compared to other metals or other aluminum series such as the 2xxx

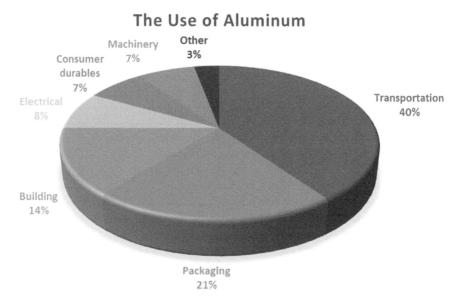

FIGURE 6.1   The use of aluminum in several sectors (Based on Woodford, 2021).

series, aluminum 6061 has lower strength (Nulhaqem & Abdul, 2013). Therefore, effort is required to make better its properties, especially its mechanical properties. Aluminum 6061 is a material widely used in the industrial world such as automotive and aerospace because it has high strength, is lightweight, and is corrosion-resistant. This alloy is commonly applied in the transportation industry, such as truck frames, rail coaches, shipbuilding operations, and military and commercial bridges (Abo Zeid, 2019).

One of the uses of aluminum alloy AA 6061 in the aerospace world is the wings and body of small-scale aircraft. Meanwhile, in the automotive industry, AA6061 is used to manufacture several types of important vehicle parts, such as wheels, panels, and even in-vehicle structures (Wardani et al., 2022). This material is widely used because it has good extrusion, formability, and weldability. In addition, this material also has medium hardness and strength, good corrosion resistance, and a good surface finish (Andoko et al., 2020).

Various processes can be applied to make better the properties of aluminum, such as mechanical and physical properties (Tsamroh et al., 2018). One of them is heat treatment. One kind of heat treatment is aging (Chacko & Nayak, 2014); this treatment can produce homogeneous and evenly distributed precipitates so that an optimal increase in the mechanical properties of the material could be obtained (Rymer et al., 2021). While the aging process is of several kinds, aging treatments that are often used to improve mechanical properties are natural aging and artificial aging (Cochard et al., 2017).

Artificial aging is the aging for aluminum alloys which are treated with age hardening in a hot state. In this study, the strength of AA6061 will be increased through by combining the treatments of natural aging and artificial aging, but this study focuses on the microstructural changes that occur during the aging process. Several studies using artificial aging methods that aim to improve the material's mechanical properties have been carried out on aluminum alloy Al 2024 at a temperature of 0°C–177 °C for 20 hours and have shown an increase in hardness (Prudhomme et al., 2018). The application of the artificial aging treatment on aluminum alloy Al 7075 with a temperature of 120°C with time variations for 60, 120, 180, and 240 minutes achieved a maximum increase in hardness at 120 minutes of holding (Lee et al., 2018). Furthermore, the Al-Si-Mg-Cu alloy with artificial aging treatment achieved maximum tensile strength at an aging temperature of 180°C with a holding time of 240 minutes (Jin et al., 2018). Aluminum alloy AA1350-H19 gained the highest tensile strength in the artificial aging treatment at a temperature of 200 °C for 4 hours of holding (Flores et al., 2018).

Based on the explanation above, it is necessary to study how changes in the microstructure of Al6061 due to natural aging and artificial aging heat treatments can affect the mechanical properties of the material. Two important parameters strongly influence changes in the microstructure of Al6061, namely, heating temperature and holding time (Tsamroh et al., 2018). The heating temperature and holding time used in the heat treatment should be determined precisely; the use of a heating temperature that is too high and holding for a long time will actually damage the material microstructure. This will also affect the mechanical properties of the material.

6.2   ALUMINUM

Materials have been classified into four major materials: metal, composite, ceramic, and fiber. Metal is divided into ferrous metals and non-hero metals. In general, aluminum can be classified into three big groups: wrought non-heat-treatable alloys, wrought heat-treatable alloys, and casting alloys (Davis, 1993). Meanwhile, aluminum casting alloys are classified into two types: alloys that have the ability to be treated by using heat treatment and alloys that do not have the ability to be treated by using heat treatment. Table 6.1 presents the aluminum classification and the aluminum alloy naming code.

Aluminum is obtained from certain types of clay (bauxite). Bauxite is first separated from pure alum (aluminum oxide). Bauxite is one of the most important materials for aluminum production, which is hydrated aluminum oxide containing 1 to 20% Fe_2O_3; 1 to 10% silicate to a lesser extent of zirconium, titanium, vanadium, and several transition metal oxides; 20 to 30% is water, and 50 to 60% Al_2O_3. Bauxite can be purified by using the process of Bayer. By filtration, sodium hydroxide can dissolve the crude bauxite and be separated from hydrated iron oxide and other insoluble foreign substances (Davis, 2001).

Then the molten aluminum oxide is calcined by an electrical procedure. Because the melting temperature of aluminum oxide is very high, namely, 2050°C, the processing of aluminum is very difficult. Aluminum metal has the symbol Al, which has a specific gravity of 2.6–2.7 with a melting point of 659°C. Aluminum is a soft metal and is harder than lead but softer than zinc. The color of aluminum is bluish-white. Aluminum can be produced through electrolysis process. The electrolysis process developed for industrial production is the Hall–Heroult electrolysis process. The process is electrolysis of alumina (Al_2O_3) solution in molten cryolite (Na_3AlF_6) at a temperature of 960°C to produce molten aluminum.

TABLE 6.1
Aluminum Classification and Naming Code

Aluminum Type	Classification	Naming Code
Aluminum alloy for machining	Wrought non-heat-treatable alloy	Pure Al (1000 series)
		Al-Mn Alloy (3000 series)
		Al-Si Alloy (4000 series)
		Al-Mg Alloy (5000 series)
	Wrought heat-treatable alloy	Al-Cu Alloy (2000 series)
		Al-Mg-Si Alloy (6000 series)
		Al-Zn Alloy (7000 series)
Aluminum alloy for casting	Non-heat-treatable casting alloy	Al-Si Alloy (Silumin)
		Al-Mg Alloy (Hydronarium)
	Heat-treatable casting alloy	Al-Cu Alloy (Lautal)
		Al-Si-Mg Alloy (Silumin, Lo-ex)

Source: Irawan, 2015.

6.2.1 PURE ALUMINUM

Aluminum is obtained in the molten state by electrolysis, which generally attains purity of 99.85% by weight. However, if further electrolysis is carried out, aluminum will be obtained with a purity of 99.99%. Corrosion resistance varies with purity; generally, 99.0% or above purity can be used in resistant air for many years. The electrical conductivity of Al is about 65% of the electrical conductivity of copper, but its density is about one-third of that of copper, so it is possible to expand its cross-section. Therefore, it can be used for cables and in various forms, such as a thin sheet (foil). Al with a purity of 99.0% can be used in this case. Al with that level of purity is used for reflectors that require high reflectivity and electrolytic coders.

6.2.2 ALUMINUM ALLOY

Aluminum alloys are grouped in various standards by various countries in the world. However, the most well-known and perfect classification is the Aluminum Association (AA) standard in America which is based on the previous standard from Aluminum Company of America (Alcoa).

Table 6.2 presents the physical characteristics of aluminum.

6.2.2.1 Al-Cu Alloy

Al-Cu and Al-Cu-Mg alloys are one of the main aluminum alloys. Copper is the main alloying element in aluminum in the 2000 series aluminum, which is often added with Mg as an additional alloying element (Davis, 2001). Al-Cu-Mg alloys contain 4% Cu and 0.5% Mg, which can harden greatly within a few days by aging at ordinary temperatures after solution heat treatment. Aluminum alloyed with Cu has poor corrosion resistance, so it is necessary to coat the surface with pure aluminum or a corrosion-resistant aluminum alloy (alclad plate). However, the alloy is used as an aircraft material (Surdia & Saito, 1999). Aluminum and copper alloys (Al-Cu alloys) are aluminum alloys known as duraluminium or super duraluminum. Duraluminum is also often referred to as duralumin or duralium (Junkers, 2014).

TABLE 6.2
Physical Characteristics of Aluminum

Character	High Pure Aluminum
Crystal structure	FCC
Density at 20°C (sat. 10^3 kg/m³)	2.698
Melting point (°C)	660.1
Wire heat creep coefficient 20°C~100°C (10^6/K)	23.9
Heat conductivity 20°C~400°C (W/(m-K))	238
Electrical resistance 20°C (10^{-8} KΩ-m)	2.69
Modulus of elasticity (GPa)	70.5
Stiffness modulus (GPa)	26.0

6.2.2.2 Al-Mn Alloy

Manganese (Mn) is the main element in 3000 series aluminum alloys. Generally, this alloy is an alloy that cannot be heat treated but has 20% more strength than 1000 series aluminum (Davis, 2001). The addition of the element Mn to aluminum can strengthen aluminum without reducing its corrosion resistance and is used to make corrosion-resistant alloys. The alloys Al-1.2% Mn and Al-1.2%Mn-1.0%Mg are alloys 3003 and 3004 that are used as corrosion-resistant alloys without heat stiffening (Surdia & Saito, 1999). This alloy is often used for kitchen utensils and panels (Irawan, 2015).

6.2.2.3 Al-Si Alloy

The main alloying element in 4000 series aluminum alloys is silicon. Silicon can be added in sufficient quantities (nearly 12%). The addition of silicon elements to the alloy lowers its melting range without causing the alloy to become brittle. This alloy is widely used in welding wire as brazing (Davis, 2001). Al-Si alloys have very good fluidity, have a good surface, have no heat flexibility, and are very good as alloy castings (silumin). This alloy also has good corrosion resistance, is very light, has a small expansion coefficient, and is a good conductor of electricity and heat. Al-12%Si alloys are widely used for cast alloys (silumin). The alloy that is treated with dissolution and aging is called silumin. The properties of silumin can be increased by applying heat treatment and slightly improved by alloying elements. Generally, alloys with 0.15–0.4% Mn and 0.5% Mg are used. Alloys that require heat treatment are added with Mg as well as Cu and Ni to provide hardness when hot; these materials are commonly used for motor pistons (Surdia & Saito, 1999).

6.2.2.4 Al-Mg Alloy

The 5000 series aluminum alloy is mainly alloyed with Mg. When combined with manganese, it produces a medium to high strength working alloy. As a hardener, magnesium is better than manganese, with 0.8%Mg equivalent to 1.25%Mn (Davis, 2001). Al-Mg alloy has good corrosion resistance; this alloy has long been called hydronalium and is known as an alloy that is resistant to corrosion (Surdia & Saito, 1999). To increase the strength of the alloy against stress corrosion, Mn and Cr elements are added (Irawan, 2015).

Alloys with a content of 2–3% Mg have the character of being easy to forge, roll, and extract. Aluminum alloy 5005 is an alloy that has a low Mg content and is often used as an accessory. Aluminum alloy 5052 is an alloy that is often used as a forging or construction material. Aluminum alloy 5056 is the strongest alloy where it is used after being hardened by strain hardening if high hardness is required. The annealed alloy 5083 is an alloy of 4.5%Mg, which is strong and easy to weld; therefore, it is used in liquefied natural gas (LNG) tanks (Irawan, 2015; Surdia & Saito, 1999).

6.2.2.5 Al-Mg-Si Alloy

Aluminum is a non-ferrous metal that is widely used in the industry. It is a light metal with a density of $2.7 g/cm^3$ and a melting point of 600°C. Aluminum has good corrosion resistance and is also a good conductor of heat and electricity. There is an oxide layer (Al_2O_3) on the aluminum surface, which serves to protect it from corrosion. This layer is hard and has a high melting point of about 2050°C. Because the melting point is much higher than that of the parent metal, the coating becomes a serious problem in the aluminum welding process, making it difficult to mix the base metal and filler metal and thus causing incomplete melting and resulting in defects in the form of fine holes in the weld (Ogunsemi et al., 2021). To remove the oxide layer (Al_2O_3), the surface to be welded must first be brushed with a steel brush. The use of noble gases during welding will prevent the oxide layer from forming again and prevent unwanted deposits from forming during the welding process.

Aluminum has light properties, good electrical, good resistance to corrosion, and heat conductivity, and is easy to form both through forming and machining processes. In nature, aluminum is a kind of oxide that is stable; thus, reduction method cannot be applied to aluminum that is usually applied in other metals. The reduction of aluminum only can be made by applying electrolysis method. In order to improve its mechanical strength, several elements can be added, such as Mg, Cu, Si, Zn, Mn, Ni, and so on, together or individually, and also to improve other good properties of aluminum, such as wear resistance, corrosion resistance, low coefficient of expansion, etc. Aluminum alloys can be divided into two groups, namely, sheet aluminum and cast aluminum. Aluminum (99.99%) has a specific gravity of 2.7 g/cm^3; above the magnesium (1.7 g/cm^3) and beryllium (1.85 g/cm^3) or about 1/3 of the specific gravity of iron or copper, a density of 2.685 kg/cm^3, and its melting point is 660°C. Aluminum has the higher strength to weight ratio compared to steel. Its electrical conductivity is 60% more than copper, so it is used for electrical equipment. In addition, aluminum is a good conductor of heat and has good reflecting properties. Therefore, it is also used in engine components, heat exchangers, reflecting mirrors, chemical industry components, etc. (Irawan, 2015). The corrosion-resistant properties of aluminum are obtained from forming an aluminum oxide layer on the aluminum surface. This oxide layer is firmly and tightly attached to the surface and is stable (does not react with the surrounding environment) to protect the inside. Aluminum and its alloys have unique properties that make aluminum one of the simplest, most economical, and often applied metallic materials for various applications and it ranks second only to steel in its use as a structural metal (Davis, 2001).

The 6xxx series aluminum alloys contain Mg and Si in the right ratios to form Mg_2Si when heat treated. Even though it is not as strong as the 2xxx and 7xxx series aluminum alloys, the 6xxx series aluminum alloys have good properties, such as weldability, formability, corrosion resistance and machinability (Andersen et al., 2018).

In addition to the above properties, 6xxx series aluminum alloy is also very good for formability for forging and extrusion and good for high formability at ordinary temperatures. After processing, these alloys can be strengthened by heat treatment (Surdia & Saito, 1999). Tables 6.3–6.5 present the chemical composition of Al6061.

6.3 PRECIPITATION HARDENING

Precipitation hardening, also known as particle hardening, is a technique in heat treatment. It is a metal alloy hardening process by spreading fine particles evenly (Ataiwi et al., 2021). The strength and hardness of the metal can be increased by the formation of very small uniformly distributed particles that occur in the second

TABLE 6.3
Chemical Composition of Al6061

Element	Value %
Silicon (Si)	0.40–0.80
Iron (Fe)	0.70
Copper (Cu)	0.15–0.40
Manganese (Mn)	0.15
Magnesium (Mg)	0.80–1.20
Chromium (Cr)	0.04–0.35
Zinc (Zn)	0.25
Titanium (Ti)	0.15
Other (Each)	0.05
Other (Total)	0.15
Aluminum (Al)	Balance

Source: ASTM B221.

TABLE 6.4
Properties of Al6061 by Heat Treatment

Alloy	Composition	Temper	Tensile Strength (MPa)	Yield Strength (MPa)	Elongation in 50 mm (%)
6061	1.0 Mg, 0.6 Si, 0.2 Cr	O	125	55	25
		T4	245	245	25
		T6	315	315	12
		T91	410	400	6

Source: Surdia & Saito, 1999.

TABLE 6.5
Mechanical Properties of Al-Mg-Si Alloy

Alloy	Condition	Tensile Strength (kgf/mm²)	Creep Strength (kgf/mm²)	Elongation (%)	Shear Strength (kgf/mm²)	Hardness (Brinell)
6061	O	12.6	5.6	30	8.4	30
	T4	24.6	14.8	28	16.9	65
	T6	31.6	28.0	15	21.0	95

Source: Surdia & Saito, 1999.

stage of the original metric phase. Examples of alloys that are increased in hardness by precipitation hardening are aluminum-copper, copper-beryllium, copper-tin, and magnesium-aluminum. Some irons can also increase hardness through precipitation hardening (precipitation), but precipitation on iron is a different phenomenon, although the healing process is almost the same (Callister & Rethwisch, 2015).

For precipitation of supersaturated solid solutions, the basic requirement of a precipitation-hardening alloy system is that the solubility limit of the solid should decrease with decreasing temperature. The hardening heat treatment procedure is first subjected to a dissolution heat treatment at a high temperature and then rapidly cooled in water or other cooling media (Bishop & Smallman, 1999).

Rapid cooling can inhibit the phase separation so that at low temperatures, the alloy is in an unstable supersaturated state, but after a rapid cooling process, if the alloy undergoes "aging" treatment for a long time, a second phase precipitates. This precipitation occurs through the process of nucleation and growth, fluctuations in the concentration of dissolved material form clusters of small atoms in the lattice, which become nucleates. As the size of the sediment becomes finer as the temperature at which precipitation is lowered, and the alloy undergoes a significant increase in hardness which is associated with a critical dispersion of the precipitate. If aging is allowed to continue at a certain temperature, there will be coarsening of the particles (small particles tend to dissolve again, and large particles get bigger). Coarser particles gradually replace numerous particles that are finely dispersed with large dispersion distances. In this state, the alloy becomes softer, and the metal is said to be in the form of late aging (ASM International Handbook, 2001).

Heat treatment improves various mechanical and alloying properties because both dissolved atoms and point defects above equilibrium concentrations are maintained during this process. The rapid cooling process often eliminates lattice strain. The property that undergoes the most change is the electrical resistance which is usually a very large increase. On the other hand, the mechanical properties are not significantly affected (Bishop & Smallman, 1999).

Changes in the properties of the quenched material after aging are more pronounced. In particular, the mechanical properties undergo major modifications. For example, the tensile strength of duralumin (an aluminum alloy – 4% copper containing magnesium silicon and manganese) can be increased from 0.21 to 0.41 GN/m^2. Structural sensitive properties such as hardness, yield stress, and so on, of course, depend on the distribution of the phase structure; thus, such alloys experience softening when the finely dispersed precipitate hardens (Bishop & Smallman, 1999). The process of precipitation hardening or hardening can be divided into several stages (Callister & Rethwisch, 2015) as follows:

1. Solution heat treatment, which is heating the alloy above the solvus line.
2. Rapid cooling (quenching).
3. Precipitation heat treatment (aging), where the workpiece is heated to a temperature of T2, where the supersaturated solid solution begins to form the phase. This phase appears in the form of a fine precipitate which is dispersed and increases the strength of the metal. The heating duration depends on the formation stage of the optimum solidifying sediment.

There are several types of precipitation heat treatment (Callister & Rethwisch, 2015), which are as follows:

a. Natural Aging

In the natural aging process, the alloy does not experience heating. Only left at room temperature, in this process, it takes a long time, and the strengthening effect given is not so great. The precipitates in the matrix are still random (Triantafyllidis et al., 2015).

b. Artificial Aging

Artificial aging is a process where the alloy is heated to a certain temperature. At this stage, the precipitates are evenly distributed and form groups; at this stage the optimum strengthening effect can be produced (Abo Zeid, 2019).

c. Over-Aging

Over-aging is the aging process which is carried out for too long, or where the temperature is too high; at this stage, the precipitate and matrix are in balance. Over-aging can reduce the strength of the material that has been achieved previously (Liao et al., 2020).

6.3.1 NATURAL AGING

Natural aging is the aging for aluminum alloys which are treated with age hardening in a cold state. Natural aging occurs at room temperature between 15°C and 25°C and with a holding time of 5 to 8 days.

6.3.2 ARTIFICIAL AGING

Artificial aging is the aging for aluminum alloys that are treated with age hardening in a hot state. Artificial aging occurs at temperatures between 100°C and 200°C and with a holding time of 1 to 24 hours (Smith et al., 2015). After solution heat treatment and quenching, alloy hardening can be achieved in two ways: at room temperature (natural aging) or by precipitation heat treatment (artificial aging). Aging at room temperature will take a longer time, usually around 96 hours, to achieve a more stable strength, whereas if aging is done artificially, the aging time depends on the heating temperature. The higher the aging temperature, the shorter the time required to reach a certain strength (Cochard et al., 2017).

The use of the alloy hardening method depends on the type of alloy that one wants to increase the hardness. For alloys with a slow precipitation reaction, precipitation is always carried out at temperatures above the room temperature (artificial aging). In contrast, natural aging is sufficient for those with a fast precipitation reaction to obtain desired mechanical properties. In the aging process, the formation and growth of nuclei occur, leading to the formation of stable precipitates. The formation of this

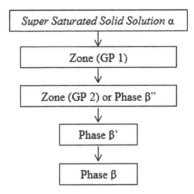

FIGURE 6.2 Phase change sequence in the artificial aging process (Based on Smith et al., 2015).

phase occurs through several phase transitions which also affect the mechanical properties of the alloy (Liao et al., 2020).

At the artificial aging stage in the age-hardening process, several variations of treatment can be carried out, which can affect the results of the age-hardening process. One of these variations is the artificial aging temperature variation. The artificial aging temperature can be set at the temperature at which the aluminum alloy crystallizes (150°C), below the crystallization temperature, or above the crystallization temperature of the aluminum alloy metal (Smith et al., 2015).

Taking the artificial aging temperature into consideration, a temperature between 100°C and 200°C will affect the level of hardness because there will be changes in the phase or structure of the artificial aging process. This phase change will contribute to the hardening of the aluminum alloy (Smith et al., 2015). The phase change in the artificial aging process is explained in Figure 6.2.

The explanation of the diagram in Figure 6.2 is as follows.

a. Supersaturated Solid Solution α

After the aluminum alloy passes through the solution heat treatment and quenching stages, a supersaturated solid solution will be obtained at room temperature. Under these conditions simultaneously, the atomic vacancies in thermal equilibrium at high temperatures remain in place. After cooling or quenching, the aluminum alloy becomes soft compared to its initial condition (Singh et al., 2018).

b. Zone (GP 1)

Zone (GP 1) is a precipitation zone formed by low aging or aging temperatures and is formed by the segregation of Mg-Si atoms in a supersaturated solid solution. Zone (GP 1) will appear in the early stage(s) of the artificial aging process. This zone is formed when the artificial aging temperature is below 100°C, and the Zone (GP 1) will not be formed at too high an artificial aging temperature. The formation of the Zone (GP 1) will begin to increase the hardness of the aluminum alloy (Smith et al., 2015).

If artificial aging is set at a temperature of 100°C, then the phase change stage is only until the formation of Zone (GP 1). The process of hardening from a supersaturated solid solution to forming a Zone (GP 1) is commonly referred to as first stage hardening (Yang et al., 2016).

c. Zone (GP 2) or β" Phase

After the artificial aging temperature passes 100°C and above, the β" or Zone (GP 2) phase will begin to appear. At a temperature of 130°C, a zone (GP 2) will be formed, and if the artificial aging holding time is fulfilled, the optimal hardness level will be obtained (Andersen et al., 2018).

Usually, the artificial aging process stops when a zone (GP 2) is formed. A fine intermediate phase is formed (β" precipitation) because after passing through the Zone (GP 2), the alloy will become soft again. If the artificial aging process continues until the β" or Zone (GP 2) phase is formed, it is called the second stage of hardening.

d. β' Phase

Suppose the aging temperature of the aluminum alloy is increased or the aging time is extended but the temperature remains constant. In that case, it will form precipitation with a stable crystal structure different from the phase. This phase is called the intermediate phase or β' phase. The formation of this β' phase can still contribute to an increase in the hardness of aluminum alloys. The increase in hardness that occurs in the β' phase is very slow (Chen et al., 2021).

e. β Phase

The holding time in artificial aging is one component that can affect the results of the overall age-hardening process. As with temperature, the holding time in the artificial aging stage will affect changes in the structure or phase of the aluminum alloys. If the temperature increases or the aging time is extended, then the β' phase changes to the β phase. If the phase is formed, it will cause the aluminum alloy to become soft again. Meanwhile, artificial aging holding time must be selected carefully (Rymer et al., 2021).

The relationship between aging time and aluminum alloy hardness begins with a phase change process that is formed in the precipitation hardening process, where the phase starts from a supersaturated solid solution after the quenching process. Then the alloy will experience aging or the appearance of new precipitates with time.

Several previous studies have developed the artificial aging method to improve the mechanical properties of aluminum alloy. Aluminum alloy AA7049 has experienced an increase in tensile strength due to multistage heat treatment (there are two times of heating). Multistage heat treatment, called the retrogression and re-aging process in this study, has been shown to increase the tensile strength of aluminum alloy AA7049 as evidenced by changes in the microstructure, where the sediment grows and fills grain boundaries (Ranganatha et al., 2013). Furthermore, previous study about the multistage-aging process on Al-Zn-Mg-Cu alloys with two heating showed a change in the microstructure, which means that the treatment influences the mechanical properties of the aluminum alloy (Mandal et al., 2020).

In 2017, the artificial aging method was developed into a multistage artificial aging method, with variations in the number of stages of aging (there are single stage, double stage, and triple stage aging) which are carried out to improve the mechanical properties of Al-Cu alloy, which through this treatment produces tensile strength and hardness. In other words, this treatment makes the Al-Cu alloy more resilient and increases its hardness (Tsamroh et al., 2017).

6.4 MICROSTRUCTURE CHANGE

Metals are generally constructed from a large number of crystals (the grains are referred to as grains of sand on a beach) consisting of one or more phases. Generally small, ranging from 10 μm to 1 μm, but there are also grains with sizes ranging from nm to cm; this microscopic metal arrangement is called a microstructure and can only be observed using a microscope. The microstructure of the grain size affects the strength of the material based on the grain size. Grain size cannot be used to control the strength of aluminum or its alloys, but it reduces the risk of hot cracking (Chen et al., 2021; García-rentería et al., 2020).

This study combines natural aging and artificial aging treatments intending to know how the microstructure changes Al6061 in natural aging treatment only with specimens receiving artificial aging treatment. Figure 6.3 shows the heat treatment diagram.

The research method used in this study is a laboratory experimental method which is intended to obtain descriptive data about changes in the microstructure of aluminum 6061 with natural aging, and natural aging treatments followed by artificial aging. Natural aging was conducted at room temperature for 7 days after the solution

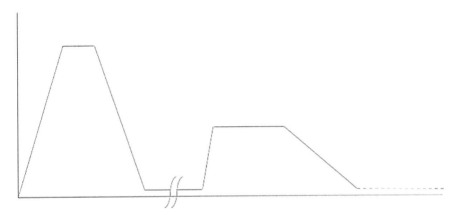

FIGURE 6.3   Heat treatment diagram.

heat treatment process at a temperature of 540°C and cooled rapidly using the mixture of water and dromus oil with a ratio of 1:1. The dependent variable of this study was microstructure change, while the independent variable of this study was the duration of holding time during the artificial aging process (2, 4, and 6 hours) with a temperature of 200°C. Figures 6.4–6.8 are the results of microstructure testing on Al6061 after getting natural–artificial aging treatment using an optical microscope. The microstructure of Al6061 was taken with a magnification of 200×. Changes in the microstructure of the material due to heat treatment can simply be seen in the grain size formed.

Precipitation that is spread evenly can increase the hardness and tensile strength of the material but causes the material to have brittle properties (Polmear, 2004). Based on the observation in Figure 6.4 it can be said that raw material of Al6061 has heterogeneous grain size and few residues on the surface of the specimen. Figure 6.5 is an image of the microstructure of Al6061 with natural aging treatment; it can be seen that the grain size is smaller than that of the raw material specimen (Figure 6.4). Figure 6.5 also shows the presence of residues formed on the surface of the specimen but not evenly distributed.

Figure 6.6 is a specimen of Al6061 with natural aging treatment followed by artificial aging treatment for 2 hours. From the figure, it can be seen that the grain size is more homogeneous, and the precipitates marked with black spots are seen more

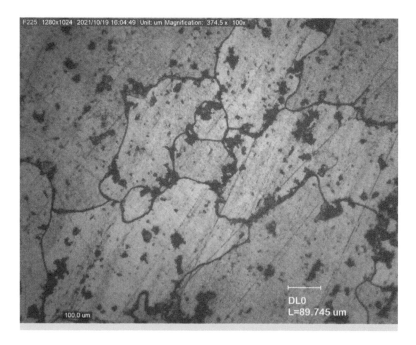

FIGURE 6.4 Microstructure of Al6061 without treatment (raw material).

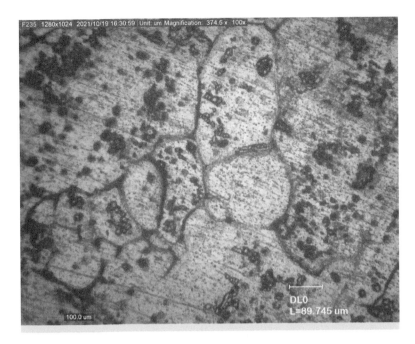

FIGURE 6.5   Microstructure of Al6061 with natural aging treatment.

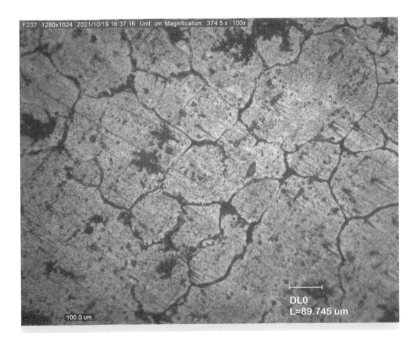

FIGURE 6.6   Microstructure of Al6061 with natural aging followed by artificial aging for 2 hours.

FIGURE 6.7   Microstructure of Al6061 with natural aging followed by artificial aging for 4 hours.

FIGURE 6.8   Microstructure of Al6061 with natural aging followed by artificial aging for 6 hours.

evenly on the surface of the specimen, both on the grainline and within the grain. Figure 6.7 is a specimen of Al6061 with natural aging treatment followed by an artificial aging process for 4 hours. The grain size of the specimen looks more homogeneous than the previous specimen, and the distribution of precipitates on the surface of the specimen also looks more even.

The last specimen shown in Figure 6.8 is a specimen with natural aging followed by artificial aging for 6 hours. In this specimen, it can be seen that the grain size has increased again; this is probably due to the material experiencing over-aging due to a longer holding time. The precipitate formed on Al6061 which is an Al-Mg-Si alloy is Mg_2Si; usually the precipitate formed is in the form of rods, needles, and laths (Andersen et al., 2018). The precipitation hardening process in Al-Mg-Si alloys can be identified in five stages:

$$\text{Supersaturated Solid Solution } \alpha \rightarrow \text{GP Zone } 1 \rightarrow \beta'' \rightarrow \beta' \rightarrow \beta = Mg_2Si$$

The hardening of the precipitate should be coherent or semi-coherent, usually the aluminum matrix has coherent or semi-coherent precipitates (S. Coriell, 2000). Coherent precipitates have a small mismatch lattice with the metal matrix, and there is a tight interfacial tension lattice. A small coherence value will affect the hardness of the metal; the smaller the coherence value, the higher the hardness of the metal will be, and the phase formed is called the β' phase, whereas if the coherence value is not there at all, the hardness of the metal will decrease (this is called the β phase). The change from phase β'' to phase β' occurs when the highest hardness number increases (Marioara et al., 2002).

This research proves that by applying artificial aging after natural aging, the β' phase can be identified which gives the material strengthening properties. It is proven by the amount of precipitate formed, which increases with increasing time with a constant temperature. In the artificial aging process, the β' phase is formed from the transition of GP zone 2 (β'') to the β' phase, which leads to coherent precipitates against the Al-Mg-Si alloy matrix. The first aging peak, the transition from GP zone 2 (β'') to the β' phase, increases significantly so that the hardness of the material in the β' phase increases. If the aging temperature or the aging time is extended but with a constant temperature, precipitation with a different crystal structure from the phase will be formed (Jin et al., 2018).

Changes in the microstructure of Al6061 in this study with natural aging and natural aging treatments followed by artificial aging were quite significant, and this can be seen from changes in grain size and the formation and distribution of precipitates (Mg_2Si). Based on Figure 6.5–Figure 6.8, the precipitate (Mg_2Si) which is the β' phase formed during the heat treatment process is shown in black spots that spread on the surface of the specimen. The precipitates formed are rods.

Based on the theory, the hardness number of a material will increase if the grain size gets smaller (Hajihashemi et al., 2016). To prove this, hardness testing was carried out on all Al6061 specimens (raw material, natural aging, and natural aging followed by artificial aging). Hardness testing was carried out using a Rockwell Hardness Tester machine, with a major load of 100 kg, and a 1/8" steel ball indenter. The test was carried out on the E scale. Based on observations of the results of the microstructure

TABLE 6.6
Hardness Number of Al6061

Specimen	Holding Time (Hour)	Hardness (HR$_E$)
Raw material	-	98
Natural aging	-	84,6
	2	93,4
Artificial aging	4	106,8
	6	105,8

test, it can be seen that the specimens with relatively small and homogeneous grain sizes are those with natural aging treatment followed by artificial aging for 4 hours. Table 6.6 presents the result of hardness testing on Al6061.

From Table 6.6, it can be observed that specimens achieved the highest hardness number with natural aging treatment, followed by artificial aging treatment with a holding time of 4 hours with a figure of 106.8 HRE. Specimens that only received natural aging treatment had the lowest hardness number, which was 84.6 HRE. While the specimens underwent natural aging treatment followed by artificial aging with a holding time of 6 hours, the hardness number slightly decreased compared to specimens with a holding time of 4 hours, which was 105.8 HRE.

When compared with the hardness number of the raw material, it can be seen that the treatment with natural aging followed by artificial aging for 4 and 6 hours showed an increase in the hardness number. Thus, it can be estimated that the most optimum treatment to increase the hardness of Al6061 is natural aging followed by artificial aging with a holding time of 4–6 hours. However, specimens that only underwent natural aging treatment had the lowest hardness numbers. Thus, it can be concluded that the natural aging treatment has no significant effect on increasing the hardness of Al6061. The results of this study are almost the same as that of studies carried out by previous researchers, where natural aging did not affect the trend of increasing specimen hardness (Wardani et al., 2022). The holding time during artificial aging certainly affects the transformation/change of the microstructure of a material, which is also related to the hardness of the material (Rymer et al., 2021).

6.5 CONCLUSION

According to the discussion presented in this chapter, the following conclusions could be drawn:

1. The microstructure change on Al6061 was obtained due to the heat treatment process. The microstructure change is mainly affected by the holding time during artificial aging.
2. The best result of this study was a specimen that had been aged artificially for 4 hours after the natural aging process, which had a smaller grain size that was likely homogeneous.

3. The change of microstructure affected the hardness of Al6061, with the highest hardness number of 106.8 HRE (specimen that aged artificially for 4 hours). The increasing hardness was thought to be caused by the formation of Mg_2Si precipitates.

ACKNOWLEDGMENT

The author would like to thank LPPM, the University of Merdeka Malang, which has provided the Internal Grant 2021.

REFERENCES

Abo Zeid, E. F. (2019). Mechanical and electrochemical characteristics of solutionized AA 6061, AA6013 and AA 5086 aluminum alloys. *Journal of Materials Research and Technology, 8*(2), 1870–1877. https://doi.org/10.1016/j.jmrt.2018.12.014

Andersen, S. J., Marioara, C. D., Friis, J., Wenner, S., & Holmestad, R. (2018). Precipitates in aluminium alloys. *Advances in Physics: X, 3*(1), 790–814. https://doi.org/10.1080/23746 149.2018.1479984

Andoko, A., Prasetya, Y. A., Puspitasari, P., & Ariestoni, T. B. (2020). The effects of artificial-aging temperature on tensile strength, hardness, micro- structure, and fault morphology in AlSiMg. *Journal of Achievement in Materials and Manufacturing Engineering. 98*(2), 49–55. https://doi.org/10.5604/01.3001.0014.1480

ASM International Handbook, V. 4. (2001). ASM handbook: heat treating. *Technology, 4,* 3470. https://doi.org/10.1016/S0026-0576(03)90166-8

Ataiwi, A. H., Dawood, J. J., & Madhloom, M. A. (2021). Effect of precipitation hardening treatments on tensile properties, impact toughness, and microstructural changes of aluminum alloy AA6061. *Materials Today: Proceedings.* https://doi.org/10.1016/J. MATPR.2021.06.011

Bishop, R. J., & Smallman, R. E. (1999). *Modern Physic Metallurgy & Materials Engineering.* Elsevier. 256.

Callister, W., & Rethwisch, D. (2015). Materials science and engineering: an introduction. *Materials Science and Engineering.* https://doi.org/10.1016/0025-5416(87)90343-0

Chacko, M., & Nayak, J. (2014). Aging behaviour of 6061 Al-15 vol% SiC composite in T4 and T6 treatments. *International Journal of Chemical, Molecular, Nuclear, Materials and Metallurgical Engineering, 8*(3), 195–198.

Chen, C., Yin, X., Liao, W., Xiang, Y., Gao, M., & Zhang, Y. (2021). Microstructure and properties of 6061/2A12 dissimilar aluminum alloy weld by laser oscillation scanning. *Journal of Materials Research and Technology, 14,* 2789–2798. https://doi.org/10.1016/ j.jmrt.2021.08.105

Cochard, A., Zhu, K., Joulié, S., Douin, J., Huez, J., Robbiola, L., Sciau, P., & Brunet, M. (2017). Natural aging on Al-Cu-Mg structural hardening alloys – investigation of two historical duralumins for aeronautics. *Materials Science and Engineering A, 690,* 259–269. https://doi.org/10.1016/j.msea.2017.03.003

Coriell, S. (2000). Precipitation hardening of metal alloys. *A Century of Excellence in Measurements, Standards, and Technology.* CRC Press 14–15.

Davis, J. R. (1993). *Aluminum and Aluminum Alloys.* ASM International.

Davis, J. R. (2001). Aluminum and Aluminum Alloys. *Light Metals and Alloys,* 66. https://doi. org/10.1361/autb2001p351

100 Nanotechnologies in Green Chemistry and Environmental Sustainability

Flores, F. U., Seidman, D. N., Dunand, D. C., & Vo, N. Q. (2018). Development of high-strength and high-electrical-conductivity aluminum alloys for power transmission conductors. In: Martin, O. (eds) *Light Metals 2018*. TMS 2018. The Minerals, Metals & Materials Series. Springer, Cham. https://doi.org/10.1007/978-3-319-72284-9_34

García-Rentería, M. A., Torres-Gonzalez, R., & Cruz-Hern, V. L. (2020). First assessment on the microstructure and mechanical properties of gtaw-gmaw hybrid welding of 6061-t6 AA. *Journal of Manufacturing Processes*, *59*(October), 658–667. https://doi.org/10.1016/j.jmapro.2020.09.069

Hajihashemi, M., Shamanian, M., & Niroumand, B. (2016). Microstructure and mechanical properties of Al-6061-T6 alloy welded by a new hybrid FSW/SSW joining process. *Science and Technology of Welding and Joining*, *21*(6), 493–503. https://doi.org/10.1080/13621718.2015.1138019

Indonesia, K. P. R. (2018). *Kemenperin Kejar Produksi Aluminium Nasional 2 Juta Ton Tahun 2025*. Kementerian Perindustrian Republik Indonesia.

Irawan, Y. S. (2015). Aluminium dan Paduannya – Material Teknik (Aluminium dan Tembaga Paduan). Chapter 12. *Material teknik. seri*, 1–8.

Jin, S., Ngai, T., Zhang, G., Zhai, T., Jia, S., & Li, L. (2018). Precipitation strengthening mechanisms during natural ageing and subsequent artificial aging in an Al-Mg-Si-Cu alloy. *Materials Science and Engineering A*, *724*, 53–59. https://doi.org/10.1016/j.msea.2018.03.006

Junkers, H. (2014). *Duralumin and the Origins of Rivets*. Duralumin and the Origins of Rivets - Airstream

Lee, Y. S., Koh, D. H., Kim, H. W., & Ahn, Y. S. (2018). Improved bake-hardening response of Al-Zn-Mg-Cu alloy through pre-aging treatment. *Scripta Materialia*, *147*, 45–49. https://doi.org/10.1016/j.scriptamat.2017.12.030

Liao, X., Kong, X., Dong, P., & Chen, K. (2020). Effect of pre-aging, over-aging and re-aging on exfoliation corrosion and electrochemical corrosion behavior of Al-Zn-Mg-Cu alloys. *Journal of Materials Science and Chemical Engineering*, *8*(2), 81–88. https://doi.org/10.4236/msce.2020.82008

Mandal, P. K., John Felix Kumar, R., & Merrin Varkey, J. (2020). Effect of artificial ageing treatment and precipitation on mechanical properties and fracture mechanism of friction stir processed $MgZn_2$ and Al_3Sc phases in aluminium alloy. *Materials Today: Proceedings*. https://doi.org/10.1016/J.MATPR.2020.10.389

Marioara, C. D., Andersen, S. J., Jansen, J., & Zandbergen, H. W. (2002). The GP-Zone to β Transformation in the Al-Mg-Si System. *Microsc. Microanal.*, *8*(2), 1444–1445.

Nulhaqem, L., & Abdul, B. I. N. (2013). *Influence of Heat Treatment on the Microstructure and Mechanical Properties of 6061 Aluminum Alloy* . Bachelor Thesis. Universiti Malaysia Pahang. June. CD7752.pdf (ump.edu.my)

Ogunsemi, B. T., Abioye, T. E., Ogedengbe, T. I., & Zuhailawati, H. (2021). A review of various improvement strategies for joint quality of AA 6061-T6 friction stir weldments. *Journal of Materials Research and Technology*, *11*, 1061–1089. https://doi.org/10.1016/j.jmrt.2021.01.070

Polmear, I. (2004). Aluminium alloys – a century of age hardening. *Materials Forum*, *28*, 1–14. https://doi.org/7F6104775CD4BCE9E9D087602166B700

Prudhomme, M., Billy, F., Alexis, J., Benoit, G., Hamon, F., Larignon, C., Odemer, G., Blanc, C., & Hénaff, G. (2018). Effect of actual and accelerated ageing on microstructure evolution and mechanical properties of a 2024-T351 aluminium alloy. *International Journal of Fatigue*, *107*(October 2017), 60–71. https://doi.org/10.1016/j.ijfatigue.2017.10.015

Puspitasari, P., Puspitasari, D., Sasongko, M. I. N., Andoko, & Suryanto, H. (2016). Tensile strength differences and type of fracture in artificial aging process of duralium against cooling media variation. *AIP Conference Proceedings*, *1778*, 0–4. https://doi.org/10.1063/1.4965746

Rajasekaran, S., Udayashankar, N. K., & Nayak, J. (2012). *T4 and T6 Treatment of 6061 Al-15 Vol. % SiC P Composite*. International Scholarly Research Network. 2012. Article ID 374719. 1–5 https://doi.org/10.5402/2012/374719

Rambabu, P., Prasad, N. E., & Kutumbarao, V. V. (2017). *Aerospace Materials and Material Technologies*. Springer. https://doi.org/10.1007/978-981-10-2143-5

Ranganatha, R., Anil Kumar, V., Nandi, V. S., Bhat, R. R., & Muralidhara, B. K. (2013). Multi-stage heat treatment of aluminum alloy AA7049. *Transactions of Nonferrous Metals Society of China (English Edition)*, *23*(6), 1570–1575. https://doi.org/10.1016/S1003-6326(13)62632-1

Rymer, L. M., Winter, L., Hockauf, K., & Lampke, T. (2021). Artificial aging time influencing the crack propagation behavior of the aluminum alloy 6060 processed by equal channel angular pressing. *Materials Science and Engineering: A*, *811*, 141039. https://doi.org/10.1016/J.MSEA.2021.141039

Singh, R., Sachan, D., Verma, R., Goel, S., Jayaganthan, R., & Kumar, A. (2018). Mechanical behavior of 304 Austenitic stainless steel processed by cryogenic rolling. *Materials Today: Proceedings*, *5*(9), 16880–16886. https://doi.org/10.1016/j.matpr.2018.04.090

Smith, W. F., Javad, H., & Prakash, R. (2015). *Introduction to Materials Science and Engineering*. McGraw-Hill Education

Surdia, T., & Saito, S. (1999). *Pengetahuan Bahan Teknik*. Pradnya Paramita. 372.

Triantafyllidis, G. K., Pukhalska, N. V., & Zagkliveris, D. I. (2015). Natural aging effects on the solutionizing heat treatment process of the A6060 Al alloy as-cast billets for profile production. *Materials Sciences and Applications*, *6*(2). *February*, 111–116.

Tsamroh, D. I. (2021). Comparison finite element analysis on duralium strength against multi-stage artificial aging process. *Archives of Materials Science and Engineering*, *109*(1), 29–34. https://doi.org/10.5604/01.3001.0015.0512

Tsamroh, D. I., Puspitasari, P., Andoko, A., Permanasari, A. A., & Setyawan, P. E. (2018). Optimization of multistage artificial aging parameters on Al-Cu alloy mechanical properties. *Journal of Achievements in Materials and Manufacturing Engineering*, *87*(2), 62–67. https://doi.org/10.5604/01.3001.0012.2828

Tsamroh, D. I., Puspitasari, P., Andoko, Sasongko, M. I. N., & Yazirin, C. (2017). Comparison study on mechanical properties single step and three step artificial aging on duralium. *AIP Conference Proceedings*, *1887*. https://doi.org/10.1063/1.5003553

Wardani, I. P., Setyowati, V. A., Suheni, I., Saputro, B., Teknik, J., Institut, M., Adhi, T., & Surabaya, T. (2022). Pengaruh natural aging Sebelum Proses artificial aging Terhadap Sifat Mekanik Aluminium 6061. *Seminar Nasional Sains Dan Teknologi Terapan VIII*, *2020*, 109–114.

Woodford, C. (2021). *Aluminum*. www.explainthatstuff.com/aluminum.html.

Yang, R. Xian, Liu, Z. Yi, Ying, P. You, Li, J. Lin, Lin, L. Hua, & Zeng, S. Min. (2016). Multistage-aging process effect on formation of GP zones and mechanical properties in Al-Zn-Mg-Cu alloy. *Transactions of Nonferrous Metals Society of China (English Edition)*, *26*(5), 1183–1190. https://doi.org/10.1016/S1003-6326(16)64221-8

7 Characterization of Self-Healing Concrete Incorporating Plastic Waste as Partial Material Substitution

Christian Hadhinata,[1] *Ananta Ardyansyah,*[2]
Viska Rinata,[2] *and M. Mirza Abdillah Pratama*[1*]
[1] Faculty of Engineering, Universitas Negeri Malang
[2] Faculty of Science, Universitas Negeri Malang
[*] Corresponding author: M. Mirza Abdillah Pratama, mirza.abdillah.ft@um.ac.id

CONTENTS

DOI: 10.1201/9781003320746-7

103

7.1   INTRODUCTION

Plastic is a type of polymer that is often used to wrap food, beverages, and children's toys. The advantageous features of plastics, such as high durability, ease of form, low density, and high strength-to-weight ratio, affect people and industry's use of plastics (Gu & Ozbakkaloglu, 2016). This is substantiated by an increase in global plastic output, which has risen from 2.3 million tons in 1950 to 448 million tons by 2015, and it is estimated that by 2050 plastic output would have more than twice (Parker, 2019).

On the other hand, the widespread production and consumption of plastics has created a new environmental threat. Only 9 percent of plastic waste has been recycled, 12 percent has been incinerated, and the remainder, up to 79 percent, has accumulated in landfills or the wild (Geyer et al., 2017). If left unchecked, plastic waste will pollute water and soil, potentially contaminating it with lead and cadmium and negatively impacting human health (Rohden et al., 2020). Based on these threats, current public concern around the world has resulted in a worldwide commitment to recyclable, reusable, and compostable industrial products. However, there is a 56 percent to 90 percent mismanagement of plastic waste in countries that produce 5 percent to 17 percent of solid waste (Geyer et al., 2017). Based on this, proper and thorough processing of plastic waste is required to reduce its negative impact.

One option is to use plastic waste as a mixture of materials on concrete. Concrete is a common building material today because of its strength under stress and ease of application. Concrete, on the other hand, is very weak to tensile, which has the capacity to develop cracks in the concrete structure (Van Tittelboom & De Belie, 2013). If the crack occurs in the structural components, the placed reinforcing bars will be harmed. Through cracking, chloride ions (Cl^-), water (H_2O), and carbon dioxide (CO_2) will enter and corrode iron (Li et al., 2020; Silva et al., 2015). Consequently, the life span of concrete and durability are decreased (Asaad et al., 2022).

Moreover, if cracks occur on construction sites with special functions, such as nuclear power plants, underground tunnels, and natural gas processing sites, it will significantly endanger both the construction itself and its workers (Safiuddin et al., 2018). On the other hand, the maintenance and repairing cost of the damaged concrete is quite expensive. Sidiq et al. (2019) estimated that the cost of concrete maintenance and repair stands at $147/m^3$, which is more expensive than its production cost of only $65 to $80/m^3$. This method is not very environmental-friendly and not economic. Therefore, a solution is needed to reduce the potential for cracking in concrete by adding natural and synthetic fibers to the concrete mixture.

Plastic waste is one of the materials that can be used as synthetic fibers in concrete that can increase the mechanical characteristics, reduce the size of the width, and the number of cracks that occur in concrete (Prisco et al., 2009; Subhashini et al., 2018). Ananthi et al. (2017) found that adding plastic waste as fiber in concrete can increase the compressive and tensile strength properties. Hidayatullah et al. (2017) utilized plastic garbage shredded into fibers on concrete, which increased compressive and tensile strength values by 9.47 percent and 39.53 percent, respectively, when compared to conventional concrete. Anandan and Alsubih (2021) also investigated the addition of plastic waste fibers up to 0.15 percent of the concrete volume and obtained the highest increase in concrete press strength in the addition of 0.15 percent plastic fiber by 33.20 N/mm^2 and higher bending strength by 5.26 N/mm^2.

Handling concrete cracks today also includes the use of self-healing materials, such as self-healing concrete. De Rooij et al. (2013) in RILEM previously divided self-healing process into two: autogenic/autogenous healing that utilizes the ability of the material itself in the healing process and autonomic healing that utilizes the addition of another material to perform the healing process. Achal et al. (2013) conducted the self-healing concrete using *strains* of *Bacillus* sp CT-5 bacteria and found that it can close cracks with the highest depth of 27.2 mm at a treatment life of 7 days. Another study conducted by Vijay and Murmu (2019) explained that cracks in the concrete were triggered by calcite particles ($CaCO_3$) as a result of the bacterial activity of *Bacillus subtilis*. Liu et al. (2020) conducted research with *Bacillus pasteurii* as a healing agent, and *recycled aggregate* as his carrier showed that the use of this bacteria could close cracks with a width of 0.28 mm and 0.32 mm compared to ordinary concrete. Based on the explanation, this chapter will explore the potential addition of plastic waste to the self-healing concrete mixture for crack prevention and closure.

7.2   PROCESSING OF PLASTIC INTO CONCRETE MATERIAL

Because polymer material is inexpensive, lightweight, and easy to maintain, it is widely employed in industry and by consumers. Polymers can be found in nature (cellulose, lignin, chitosan, protein, pectin, etc.) or can be synthesized through the industry as thermosets or thermoplastics widely used in everyday life. One of the most widely found polymers in everyday life is plastic. The wide application of plastic causes this material to be widely utilized. However, plastic has become a serious problem for the environment, along with its use (Li et al., 2021). Plastic is a waste that is difficult to degrade naturally. Therefore, reprocessing plastic into raw materials is widely done.

Polymer reprocessing is followed by degradation of macromolecules to reduce the viscosity, melt strength, and mechanical characteristics considered in recycling (Brachet et al., 2008). Plastics that are one of the macromolecules require several stages before reuse. Processing plastic into raw materials for the second time starts with collection, grouping, milling, washing, and drying. Then the plastic is converted into plastic pellets, powder, or flakes. Some of the developments of such processes include automatic separation, flotation, and density separation. After becoming a plastic seed, the printing process is carried out (Yin et al., 2015). There are two methods followed in plastic processing, especially for the polythene terephthalate (PET) type, namely mechanical and chemical processes. Mechanically, PET goes through the process of collection, sorting, washing, and destruction. The process mechanically obtains 84 percent of plastics (Ragaert et al., 2017). Chemically, the plastic processing process breaks the polymer chain which is known as depolymerization (Al-Sabagh et al., 2016). Although the process is quite long, plastic processing is the best choice in dealing with plastic waste.

The potential of processing plastic waste as fiber is widely studied today. Some studies have shown that recycled polymers can be converted into fibers with a diameter of 100 nm to 10 µm (Chinchillas-Chinchillas et al., 2020). The raw material for the production of plastic fibers can be obtained from generally available sources. Plastic is converted into fiber through a series of chemical reactions. Electrospinning is one of the approaches that can be employed in such modifications (Zander et al., 2015). Furthermore, the fiber may be converted into nanofibers, which offer a broader range of uses. When plastic fibers are turned into nanofibers, the strength, strain, and heat resistance created can rise. Reprocessing plastic into fiber can result in goods with high strength, durability, quality, and sustainability.

Plastic garbage is also commonly applied in the building industry as cement composites, asphalt, door panels, and bulkheads (Awoyera & Adesina, 2020). However, the potential utilization of plastic fiber in buildings, especially concrete production, is increasingly developed. Fiber-reinforced concrete (FRC) is a composite material that has undergone strengthening in the strain of cracks resulting from the strengthening of the fibers connecting it (Prisco et al., 2009). Fiber with high modulus can be added partially or entirely to a concrete dough. In this case, the application of plastic fibers derived from plastic waste can be used as concrete composites.

7.3   FEASIBILITY OF PLASTIC WASTE ON CONCRETE

Concrete construction materials are still being produced as the primary basic component of today's structures. The addition of various components to concrete, such as iron, glass, and polymer fibers, improves the strength of the existing concrete. Furthermore, the creation of such composites attempts to enhance strain strength, flexibility, and energy absorption (Kim et al., 2010). Plastic is produced from these components and is a commonly used material.

The use of plastic in concrete production continues to be developed today. In general, the use of plastic obtained from plastic waste in concrete production can positively impact the environment. The amount of plastic waste can be reduced and it can directly reduce the use of natural raw materials. The concrete mixture is often made by digging stones and crushing them. Excavation can cause changes in the geology of the area, whereas destruction can generate dust, which can pollute the ecosystem (Khajuria & Sharma, 2019). Therefore, the use of waste, such as non-degradable plastics as an addition to concrete, can positively impact the environment. As a result, *plastic fiber-reinforced concrete* (PRFC), in the form of concrete made by adding plastic to concrete, is a complement to the mixture (Chowdhury et al., 2013).

Plastic additions to concrete can take several forms, including aggregates, particles, and fibers (Gu & Ozbakkaloglu, 2016). Plastic aggregates are obtained from the breakdown of plastic trash. The plastic aggregate is then mixed into the concrete dough to a certain extent. The addition improves abrasion resistance, fatigue resistance, vibration absorption, and conductivity (Jaivignesh & Sofi, 2017). Plastic gets crushed into microscopic components, such as sand, when it is in the form of particles (Marzouk et al., 2005). This plastic may then be utilized in the production process to replace the rough and smooth concrete mixtures. Plastic, because of the form of the particles, can fill parts of concrete that are normally empty and increase strain. However, the interaction between plastic particles and low cement frequently requires plastic particles to be improved before they can be employed in concrete production (Yang et al., 2015).

The use of plastic in the form of fiber in the manufacturing of concrete is the next step. The conversion of plastic to fiber can have a positive impact on the properties of concrete. Plastics, as opposed to aggregates or particles in the form of fibers, can keep their flexibility and give links to one another (Simon & Milad, 2019). Nevertheless, this plastic fiber might be used in place of the steel fiber that is normally used.

Plastic fiber provides benefits over steel fiber, which is currently frequently utilized. Aside from the lower cost, plastic fiber may be better for the environment. Plastic fibers can also help avoid explosive spalling in fires by melting and creating a vacuum for expansion (Paul et al., 2020). The usage of fiber also has an effect on the strain created in the concrete manufacturing process. Fiber can help to strengthen the resultant strain and improve concrete ductility. Plastic as a concrete fiber can produce positive performance in terms of workability, compressive strength, splitting tensile strength, and flexural behavior (Ming et al., 2021).

Plastic is blended with natural resources in a certain proportion. PET plastic waste is commonly discovered and utilized in concrete mixtures. This type of plastic is commonly used in beverage and bottled water containers (Baldenebro-Lopez, 2014). PET plastic waste is converted into fiber and used in the manufacture of PRFC. Other polymers used to improve the properties of concrete include nylon, aramid, polypropylene, polyethylene, and polyester (Megasari et al., 2016). However, because this form of plastic trash is so common, PET is still the most common.

The addition of plastic to concrete impacts several characteristics of concrete, such as physical, mechanical, and durability properties. Because of the lack of bonding between PET plastic fiber and cement, the resultant concrete will have high flexibility and strength but low resistance (Adda & Slimane, 2019). However, plastic flexibility provides flexible gluing to concrete which gives an advantage to elasticity and strength. As a result, it will avoid deformation (Anandan & Alsubih, 2021). Furthermore, effective homogeneous mixing can create a tension zone, increasing the flexibility of the concrete. Plastic waste strengthens concrete resistance to tension by enhancing crack formation. It will significantly increase their density. The result concluded that concrete has a strength of 40.3 percent, with the plastic mixture having a strength of 0.9 percent. Meanwhile, the results of the split tensile test showed an increase of 54.8 percent (Ananthi et al., 2017). The composition of the plastic will affect the strengthening efficiency of the concrete matrix and cracks that have an impact on the maximum pressure can be held accountable. Physically, concrete with plastic is lighter due to reduced density, absorbs less water due to plastic substitution, and has a higher heat resistance (Fraternali et al., 2011; Zulkernain et al., 2021).

7.4 BIO-BASED SELF-HEALING CONCRETE APPEARANCE

7.4.1 SELF-HEALING CONCRETE IN ADVANCE

Self-healing concrete is defined as any process from a concrete material that can recover cracks and potentially improve performance after previous actions that have reduced the performance of the concrete constituent materials (de Rooij et al., 2013). There are two widely recognized self-healing concrete approaches: autogenous and autonomous healing, as shown in Figure 7.1.

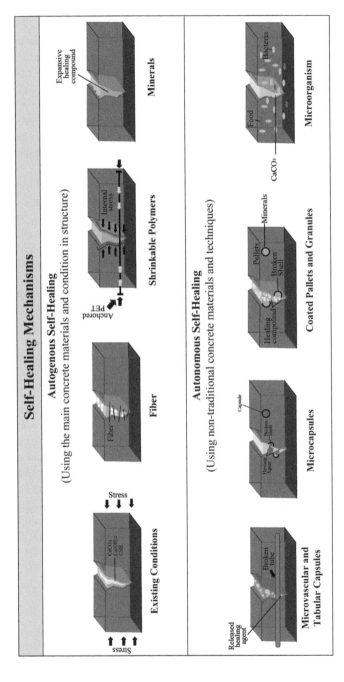

FIGURE 7.1   Self-healing concrete mechanisms (Reproduced and modified from Qureshi & Al-Tabbaa, 2020).

Qureshi and Al-Tabbaa (2020) define autogenous self-healing as the healing ability of concrete obtained from the intrinsic process of the main ingredients that make up the concrete, such as cement that undergoes a hydration process from cement that has not been adequately hydrated. Unfortunately, this mechanism results in very limited crack closure and is only effective in closing crack widths up to 50 m to 150 m. The addition of fibers and the use of superplasticizers in engineered cementitious composites (EEC) can also increase the performance of autonomous healing by limiting crack opening in concrete. PET tendons, which can shrink when activated with a heating system of concrete structures, can also compress and seal cracks. A considerable improvement in autogenous healing performance can also be achieved by using optimal cementitious additives and intelligent expansive minerals.

In contrast, autonomous self-healing is one of the healing processes that involves additional ingredients in the form of healing agents that come from other than concrete ingredients (De Rooij et al., 2013). This mechanism uses a healing agent modified and molded into an encapsulation or a vascular network made of glasses and polymers. Usually, in several studies, the healing agent used is in the form of epoxy resin, methyl methacrylate, hydrogel, and bacteria microorganisms.

7.4.2 EVALUATING TECHNIQUES USED TO VERIFY HEALING PROCESS

As illustrated in Figure 7.2, the evaluation technique used to verify the healing process in concrete is divided into three steps (de Rooij et al., 2013). The first stage is a technique for evaluating crack closure results using X-ray diffraction (XRD) and Raman spectroscopy.

The second stage involves evaluating the durability characteristics after the crack has closed. Several methods have been applied at this stage, such as permeability properties (water, air, and osmosis), absorption of water capillarity, analysis with resonant frequency, ultrasonic measurements, electrochemical and ultrasonic measurements, and resistance of concrete against steel corrosion.

The final approach is an evaluation technique that is used to verify the mechanical properties after healing recovery. At this stage, it is divided into three different

Self-Healing Evaluation Techniques		
Techniques Used to Examine Crack Healing	1. Microscopy 2. X-Ray Diffraction (XRD) 3. Raman Spectroscopy	
Techniques Used to Verify The Durability Characteristics after Crack Healing	1. Permeability 2. Capillary Water Absorption 3. Resonant Frequency Analysis 4. Ultrasonic Measurements 5. Electrochemical Impedance Measurements	6. Computed Tomography 7. Resistance against Corrosion
Techniques Used to Verify Mechanical Properties after Healing Recovery	1. Regain in Strength and Stiffness 2. Fatigue Resistance 3. Acoustic Emission Analysis	

FIGURE 7.2 Self-healing evaluation techniques.

approaches that have previously been used in research, such as regaining strength and stiffness by comparing with conventional control concrete specimens, fatigue resistance, and acoustic emission analysis.

7.4.3 MICROORGANISMS ON SELF-HEALING CONCRETE

Microorganisms in self-healing concrete or commonly called bio-concrete are one of the products that can close cracks in concrete by producing mineral compounds through microbial activity in concrete. This process is one of the autonomous healing mechanisms that can increase structural durability by reducing concrete cracks (Vekariya & Pitroda, 2013). With the help of water vapor and CO_2 in the air, the cracks will be able to recombine. Thereby, it will reduce the risk of more severe damage.

Based on microbiological studies, the application of bacteria in concrete self-healing mechanisms is natural. Although, in general, the concrete matrix seems inhospitable to all kinds of life because it is classified as a very dry and very alkaline environment, this condition is comparable to the natural system where bacteria multiply. In rocks, even at depths of more than 1 km within the Earth's crust, in deserts and ultra-wet environments, active bacteria can still be found (Atouguia, 2018). In this process, bacteria with drought and/or alkali resistance usually form spores, with a special capacity to withstand high pressures chemically and mechanically (Romano-Armada et al., 2020). Then another characteristic of spores is a low metabolic activity with a very long-life span. In addition, a small percentage of species can form spores that can survive for 200 years (European Patent Office, 2015).

This mechanism can control the width of cracks in concrete because ordinary concrete tends to form cracks in both short-term and long-term installations (Hussain Jakhrani et al., 2019). In general, the self-healing mechanism is played by certain microbes inserted into concrete to produce calcium carbonate that blocks microcracks and pores in concrete (Jena et al., 2020). Concrete supplementation with healing *agents* in stimulating crack repair is referred to as self-healing engineering (Shah & Huseien, 2020).

7.4.4 CRITERIA FOR BACTERIA IN SELF-HEALING CONCRETE

Bacteria utilized in the self-healing process must have two distinct characteristics: First, the bacteria must be able to thrive in the alkaline environment (pH12.8) caused by the water/cement combination. Second, even under harsh environmental circumstances, bacterium spores must be created (Snoeck, 2015). One form of bacterium that meets both criteria is *Bacillus* genus bacteria. *Bacillus* bacteria have a thick cell wall and are classed as gram-positive bacteria. As a result, spores may live in very alkaline environments. These spores might germinate and develop into vegetative active bacterial cells if environmental conditions are suitable (the availability of water, nutrients, and oxygen) (Kim et al., 2012).

The capacity of bacteria to induce carbonate precipitation determines the type of bacteria employed in self-healing concrete. If the circumstances are favorable, most microorganisms could induce $CaCO_3$ deposition. However, the deposition method of $CaCO_3$ bacteria differs depending on the metabolic process. The types of bacteria that have been utilized for applications on bacterium-based concrete self-healing are explained below (De Belie et al., 2018).

7.4.4.1 Bacteria Involving Nitrogen Cycles through Urea Degradation (Ureolytic Strain)

Bacteria that are commonly used in cement-based products belong to the ureolytic bacterial strain, which is highly resistant to extreme alkaline environmental conditions. These bacteria can degrade urea into ammonium ions and carbohydrates. $CaCO_3$ can occur in the presence of Ca^{2+} ions in the air, as shown by Equations 7.1 and 7.2.

$$CO(NH_2)_2\, 2H_2O + \xrightarrow{\text{Bacterial urease}} 2NH_4 + + CO_3^{2-} \qquad (7.1)$$

$$Ca_2+ + CO_3^{2-} \rightarrow CaCO_3 \qquad (7.2)$$

This group of bacteria includes many species of the genera *Sporosarcina* sp. and *Bacillus* sp. These are the species *Sporosarcina pasteurii, Sporosarcina ureae, B. pasteurii,* and *Bacillus sphaericus* (Jonkers et al., 2010). In theory, the development of 1 mole of $CaCO_3$ requires the addition of 1 mole of urea. However, this is not always the case. This is due to the fact that biochemical reactions depend extensively on the urease enzyme produced by bacteria. The productivity of carbonate production can be stated to be strongly dependent on the ureolytic activity of bacteria in concrete cracks (De Belie et al., 2018).

7.4.4.2 Bacteria Involving the Nitrogen Cycle by Assimilating from Nitrates

The other bacteria group is a group of bacteria that can create $CaCO_3$ via the nitrate reduction process. *Pseudomonas denitrificans* and *Castellaniella denitrificans* are the bacteria in this category (Karatas, 2008). This kind of bacterium is extremely resistant to harsh circumstances (such as a lack of oxygen) and is ideal for use in self-healing concrete mixes (Erşan et al., 2015). As demonstrated in Equation 7.3, the group of denitrificant bacteria utilized nitrate molecules (NO_x) as an alternate electron acceptor for the oxidation of organic carbon and formed CO_3^{2-} and HCO_{3-} ions necessary for $CaCO_3$ precipitation.

$$5HCOO^- + 2NO_3 \rightarrow N2 + 3HCO_3^- + 2CO_3^{2-} + H_2O \qquad (7.3)$$

7.4.4.3 Bacteria Involving Carbon Cycle through Oxidation of Organic Carbon

Finally, other types of bacteria widely used for self-healing concrete are some strains of alkaliphilic bacteria, which can degrade organic compounds into CO_2 and H_2O. In a high pH environment (such as concrete dough), CO_2 can be easily converted to CO_3^{2-} and in the presence of Ca^{2+}, $CaCO_3$ can be formed. Typical strains in this group are *Bacillus cohnii, Bacillus pseudofirmus,* and *Bacillus alkalinitrilicus* (Jonkers et al., 2010; Wiktor & Jonkers, 2011). In this case, calcium lactate is used as a source of organic carbon due to its less negative effect on the mechanical properties of concrete. Therefore, the overall biochemical reaction can be as follows (Equations 7.4 and 7.5):

$$Ca(C_3H_5O_3) + 6O_2 \rightarrow CaCO_3 + 5CO_2 + 5H_2O \qquad (7.4)$$

$$5CO_2 + 5Ca(OH)_2 \rightarrow 5CaCO_3 + 5H_2O \qquad (7.5)$$

From the three groups of bacteria above, it can be concluded that the strain of the ureolytic strain bacteria group had the highest carbonate yield compared to the other two types of bacteria. Based on this, it is now easier to find research that focuses on the use of ureolytic strains in concrete. However, we must be aware of ammonia production, which can cause environmental problems. Even with lower $CaCO_3$ productivity, aerobic oxidation of organic carbon produces harmless by-products. Such anoxic organic carbon oxidation can also be used under oxygen-poor conditions, eliminating the disadvantages of ureolytic strains.

7.5 THE POTENTIAL APPEARANCE OF INCORPORATING PLASTIC WASTE ON SELF-HEALING CONCRETE

The use of fiber in concrete is also one of the methods commonly used as a healing agent in self-healing concrete (Huseien et al., 2022). Several studies have also stated that the addition of fiber in concrete has advantages in terms of preventing cracking. In his research, Kwon et al. (2013) used steel fiber into concrete which showed a downward trend in the level of water permeability when the concrete was treated. This event was caused by the presence of precipitated crystalline particles formed from concrete added with fiber.

The other study was conducted by Nishiwaki et al. (2012), where the authors investigated the healing capacity of concrete by using various types of synthetic fibers. Polypropylene, ethylene-vinyl alcohol copolymer (EVOH), polyacetal (POM), and polyvinyl alcohol (PVA) were employed in the investigation. The fiber is separated into two shapes: circular and distorted shapes. According to the findings of this investigation, the addition of PVA can restore the water permeability of concrete through a self-healing process for fracture widths of up to 0.3 mm.

The usage of steel fibers and synthetic polymers from manufacturers, on the other hand, has environmental drawbacks. Adding plastic trash provides a better, safer, and more efficient environment than steel fiber, which requires a lot of energy, costs a lot of money, and is prone to corrosion (Chindaprasirt & Rukzon, 2008; JPNN. com, 2021).

Furthermore, the use of self-healing concrete with bacteria as a healing agent can significantly increase the compressive strength compared to conventional concrete. Gandhimathi and Suji (2015) used a healing agent in the form of *B. sphaericus* bacteria with a cube specimen of 150 mm × 150 mm × 150 mm and found that it can increase the compressive strength by 14.92 percent at 28 days. Bashir et al. (2016) also conducted a research about the use of bacteria from the genus *Bacillus* to increase the compressive strength of concrete at the age of 7 and 28 days. The addition of *B. subtilis* spores, calcium acetate, and urea in fly ash medium can also increase the compressive strength of concrete by 25.38 percent, 21.40 percent, and 17.97 percent after 3, 7, and 28 days, respectively (Nugroho et al., 2015). Other materials researched by Nasim et al. (2020) identified that the addition of PVA fiber, fly ash, and crystalline admixture was able to increase the compressive strength by 17.61 percent, 22.79 percent, and 36.6 percent, respectively, as compared to cracked control concrete.

Several research results revealed that the application of self-healing concrete can increase tensile strength. The study of Andrew et al. (2012) found that 5 percent microbial concentration can improve the tensile strength of concrete by 25 percent. Monishaa and Nishanthi (2008) identified an increase in tensile strength of 21.4 percent on self-healing concrete with a bacterial composition of 10^5 cells/mm and 0.4 percent of polyethylene fiber. The use of rubber particles and bacterial spores as a healing agent can also improve the tensile strength of 9.4–17.5 percent than the control concrete (Xu et al., 2019).

The addition of *B. subtilis* can increase the flexural strength after the healing process by 0.34 MPa in 28 days (Feng et al., 2021). Monishaa and Nishanthi (2008) also concluded that bacteria-based concrete and polyethylene fiber could improve flexural strength by 16.04 percent. However, the opposite result was shown by the Muslikin and Masagala's (2020) research in bacterial concrete, which proved a decrease in flexural strength from 7.80 MPa to 6.89 MPa.

Unfortunately, at this time, there have not been many studies that combine plastic fibers into self-healing concrete based on microorganisms. Reflecting on the previous explanation, these two materials will have high-profit potential if they can be applied, especially if applied on a large scale.

7.6   FUTURE SCOPE OF SELF-HEALING CONCRETE INCORPORATING PLASTIC WASTE

7.6.1   Future Environmental Development

The use of self-healing concrete with plastic waste has a very positive impact on the environment. This is because the two materials are included in sustainable building materials with the following characteristics (Stanaszek-Tomal, 2020):

• Reduction of energy consumption.
• Reduce the negative impact on the environment.
• Harmless to human health.

If the concrete has not been damaged, self-healing ability is assumed to occur. When the internal stress on the material surpasses the expected amount, the healing agent is activated (Stanaszek-Tomal, 2020).

On the other hand, processing plastic waste is one of the hot topics to be discussed and researched, especially its use for construction materials. In general, plastic waste can be processed through four methods, as shown in Table 7.1. The utilization of plastic waste in the production of self-healing concrete can certainly help reduce the accumulation of plastic waste on land, sea, and air (Nkwachukwu et al., 2013).

Material recycling methods can be used to generate self-healing concrete with the addition of plastic waste; in self-healing concrete, plastic waste can be cut into small pieces through a machine to form fibers with different shapes and sizes according to the needs and the required criteria. The method of material recycling was chosen.

TABLE 7.1
Recycling Methods for Waste Plastics

Recycling Methods	Description
Materials recycling	Using used plastic as raw material for new molds.
Thermal recycling	Burning plastic waste using heat energy.
Feedstock recycling	Thermally or chemically decomposing plastic waste for use as a reducing furnace in blast furnaces.
Chemical recycling	Decompose plastic waste thermally or chemically to be used as petrochemical raw materials.

Source: Grigore, 2017.

After all, it is thought to be the most effective way because it consumes less energy and has a lower environmental impact (Zhu et al., 2022). This concept can also recycle plastic waste, which is suitable for sustainable and modern waste processing.

The use of self-healing concrete with a mixture of plastic waste has opportunities to improve the quality of building materials. However, as mentioned by Saifee et al. (2015), the utilization of these materials, particularly microbes, has several disadvantages. The utilization of self-healing concrete with a plastic waste mixture has the potential to improve the quality of building materials. It should be noted, however, that the use of these materials, particularly bacteria, has some drawbacks; as described in Stanaszek-Tomal (2020), commonly used bacterial species (such as *B. sphaericus, B. subtilis, B. pasteurii,* and *Bacillus lexus*) have no negative impact on human health and instead have a higher ability to precipitate calcite.

7.6.2 FUTURE CONSTRUCTION DEVELOPMENT

Incorporating plastic waste on concrete has become a model material widely expected as an advanced construction material. Self-healing concrete has a lot of potential in its development to play a role in critical projects including the construction of roads and bridges, monumental buildings, erosion prevention buildings, and buildings in earthquake-prone areas (Schlangen & Sangadji, 2013). The usage of self-healing concrete made from plastic waste meets a variety of practical and economic needs.

The use of self-healing concrete for building roads and bridges is very appropriate, especially on flooded roads and bridges that are often hit by river currents. Roads or bridges do not require as much maintenance costs as conventional concrete with self-healing capabilities (Dowding, 2015). The concept of building cheap and durable roads using self-healing concrete based on plastic waste also increases the opportunity for the addition or expansion of production roads that will connect more areas to facilitate economic access.

In addition, self-healing concrete material can play a role in the construction of buildings to prevent erosion. This is very important in reducing the risk of flooding and other adverse effects of erosion, such as reducing the fertility of the surrounding soil. Then, self-healing concrete can also minimize the frequency of maintenance and renovations on monumental buildings so that the authenticity of the building can last longer.

Furthermore, self-healing concrete can be used to build houses in earthquake-prone areas. Areas with a high level of earthquake sensitivity have the potential to inflict concrete damage in the form of cracks, resulting in the destruction of a structure. With a self-healing mechanism, these cracks can be minimized and even repaired (Pinto & Brasileiro, 2021).

The use of plastic waste as a component in the production of material self-healing concrete is evolutionary. In this instance, it is possible to assert that plastic trash can be turned into useful and cost-effective building materials, and that it has critical worth for the advancement of modern construction. However, this processing model still requires in-depth research and the implementation of a more advanced development idea that includes the government as a regulator and facilitator, as well as the community as an active control and implementer.

7.7 CONCLUSION

The incorporation of plastic garbage into self-healing concrete is a construction material innovation that has been shown to increase the mechanical and durability properties of concrete. Plastic garbage can surely be utilized as an alternative to waste treatment and can help to reduce the volume of waste in the environment. This material can also be employed in difficult environmental situations, such as earthquake-prone areas, tunnels, and special-purpose buildings (e.g., nuclear power plants). To encourage the use of this material, it is hoped that in the future there will be a large amount of study that would explore the combination of these two types of materials, beginning with basic research and continuing to apply research and development.

REFERENCES

Achal, V., Mukerjee, A., & Sudhakara Reddy, M. (2013). Biogenic treatment improves the durability and remediates the cracks of concrete structures. *Construction and Building Materials*, *48*, 1–5. https://doi.org/10.1016/j.conbuildmat.2013.06.061

Adda, H. M., & Slimane, M. (2019). Study of concretes reinforced by plastic fibers based on local materials. *International Journal of Engineering Research in Africa*, *42*(January), 100–108. https://doi.org/10.4028/www.scientific.net/JERA.42.100

Al-Sabagh, A. M., Yehia, F. Z., Eshaq, G., Rabie, A. M., & ElMetwally, A. E. (2016). Greener routes for recycling of polyethylene terephthalate. *Egyptian Journal of Petroleum*, *25*(1), 53–64. https://doi.org/10.1016/j.ejpe.2015.03.001

Anandan, S., & Alsubih, M. (2021). Mechanical strength characterization of plastic fiber reinforced cement concrete composites. *Applied Sciences*, *11*(2), 852. https://doi.org/10.3390/app11020852

Ananthi, A., Eniyan, A. J. T., & Venkatesh, S. (2017). Utilization of waste plastics as a fiber in concrete. *International Journal of Concrete Technology*, *3*(January), 5. www.researchgate.net/publication/321049688%0AUtilization

Andrew, T. C. S., Syahrizal, I. I., & Jamaluddin, M. Y. (2012). Effective microorganisms for concrete (EMC) admixture—its effects to the mechanical properties of concrete. *Caspian Journal of Applied Sciences Research*, August, 419–426.

Asaad, M. A., Huseien, G. F., Memon, R. P., Ghoshal, S. K., Mohammadhosseini, H., & Alyousef, R. (2022). Enduring performance of alkali-activated mortars with metakaolin as granulated blast furnace slag replacement. *Case Studies in Construction Materials, 16,* e00845. https://doi.org/10.1016/j.cscm.2021.e00845

Atouguia, F. S. (2018). Physiology of the polyextremophile Natranaerobaculum magadiense [Universidade Nova de Lisboa]. http://hdl.handle.net/10362/55071

Awoyera, P. O., & Adesina, A. (2020). Plastic wastes to construction products: status, limitations and future perspective. *Case Studies in Construction Materials, 12,* e00330. https://doi.org/10.1016/j.cscm.2020.e00330

Baldenebro-Lopez, F. J. (2014). Influence of continuous plastic fibers reinforcement arrangement in concrete strengthened. *IOSR Journal of Engineering, 4*(4), 15–23. https://doi.org/10.9790/3021-04411523

Bashir, J., Kathwari, I., Tiwary, A., & Singh, K. (2016). Bio concrete – the self-healing concrete. *Indian Journal of Science and Technology, 9*(1), 1–5. https://doi.org/10.17485/ijst/2016/v9i47/105252

Brachet, P., Høydal, L. T., Hinrichsen, E. L., & Melum, F. (2008). Modification of mechanical properties of recycled polypropylene from post-consumer containers. *Waste Management, 28*(12), 2456–2464. https://doi.org/10.1016/j.wasman.2007.10.021

Chinchillas-Chinchillas, M. J., Gaxiola, A., Alvarado-Beltrán, C. G., Orozco-Carmona, V. M., Pellegrini-Cervantes, M. J., Rodríguez-Rodríguez, M., & Castro-Beltrán, A. (2020). A new application of recycled-PET/PAN composite nanofibers to cement–based materials. *Journal of Cleaner Production, 252,* 119827. https://doi.org/10.1016/j.jclepro.2019.119827

Chindaprasirt, P., & Rukzon, S. (2008). Strength, porosity and corrosion resistance of ternary blend Portland cement, rice husk ash and fly ash mortar. *Construction and Building Materials, 22*(8), 1601–1606. https://doi.org/10.1016/j.conbuildmat.2007.06.010

Chowdhury, S., Maniar, A. T., & Suganya, O. (2013). Polyethylene terephthalate (PET) waste as building solution. *International Journal of Chemical, Environmental & Biological Sciences (IJCEBS), 1*(2), 2320–4087.

De Belie, N., Wang, J., Bundur, Z. B., & Paine, K. (2018). Bacteria-based concrete. In F. Pacheco-Torgal, Robert E. Melchers, Xianming Shi, Nele De Belie, Kim Van Tittelboom, & Andrés Sáez (eds) *Eco-efficient Repair and Rehabilitation of Concrete Infrastructures* (Nomor 2016). Elsevier Ltd., 531–567. https://doi.org/10.1016/B978-0-08-102181-1.00019-8

De Rooij, M. R., Schlangen, E., & Joseph, C. (2013). Introduction. In de Rooij, M., Van Tittelboom, K., De Belie, N., Schlangen, E. (eds) *Self-Healing Phenomena in Cement-Based Materials*. RILEM State-of-the-Art Reports, vol 11. Springer, Dordrecht. . https://doi.org/10.1007/978-94-007-6624-2_1

Di Prisco, M., Plizzari, G., & Vandewalle, L. (2009). Fibre reinforced concrete: New design perspectives. *Materials and Structures/Materiaux et Constructions, 42*(9), 1261–1281. https://doi.org/10.1617/s11527-009-9529-4

Dowding, C. (2015). Replacement bridges for low traffic volume roads part one: Concrete hollowcore bridge. https://na.eventscloud.com/file_uploads/f85f2cdf128b930c28cc1bd21702eeae_ReplacementBridgesforLowTrafficVolumeRoads-PartOne-HollowcoreBridges.pdf

Erşan, Y. Ç., Belie, N. de, & Boon, N. (2015). Microbially induced $CaCO_3$ precipitation through denitrification: an optimization study in minimal nutrient environment. *Biochemical Engineering Journal, 101,* 108–118. https://doi.org/10.1016/j.bej.2015.05.006

European Patent Office. (2015). *"Bio-concrete" set to revolutionise the building industry: Dutch inventor of self-healing concrete named finalist for European Inventor Award.* www.epo.org/news-events/press/releases/archive/2015/20150421i.html

Feng, J., Chen, B., Sun, W., & Wang, Y. (2021). Microbial induced calcium carbonate precipitation study using Bacillus subtilis with application to self-healing concrete preparation and characterization. *Construction and Building Materials, 280,* 122460. https://doi.org/10.1016/j.conbuildmat.2021.122460

Fraternali, F., Ciancia, V., Chechile, R., Rizzano, G., Feo, L., & Incarnato, L. (2011). Experimental study of the thermo-mechanical properties of recycled PET fiber-reinforced concrete. *Composite Structures, 93*(9), 2368–2374. https://doi.org/10.1016/j.compstruct.2011.03.025

Gandhimathi, A., & Suji, D. (2015). Studies on the development of eco-friendly self-healing concrete—a green building concept. *Nature Environment and Pollution Technology: An International Quarterly Scientific Journal, 14,* 639–644. www.neptjournal.com

Geyer, R., Jambeck, J. R., & Law, K. L. (2017). Production, use, and fate of all plastics ever made. *Science Advances, 3*(7). https://doi.org/10.1126/sciadv.1700782

Grigore, M. (2017). Methods of recycling, properties and applications of recycled thermoplastic polymers. *Recycling, 2*(4), 24. https://doi.org/10.3390/recycling2040024

Gu, L., & Ozbakkaloglu, T. (2016). Use of recycled plastics in concrete: a critical review. *Waste Management, 51,* 19–42. https://doi.org/10.1016/j.wasman.2016.03.005

Hidayatullah, S., Kurniawandy, A., & Ermiyanti. (2017). Pemanfaatan Limbah Botol Plastik Sebagai Bahan Serat Pada Beton. *Jom FTEKNIK, 4*(1), 1–7.

Huseien, G. F., Nehdi, M. L., Faridmehr, I., Ghoshal, S. K., Hamzah, H. K., Benjeddou, O., & Alrshoudi, F. (2022). Smart bio-agents-activated sustainable self-healing cementitious materials: an all-inclusive overview on progress, benefits and challenges. *Sustainability (Switzerland), 14*(4). https://doi.org/10.3390/su14041980

Hussain Jakhrani, S., Qudoos, A., Gi Kim, H., Kyu Jeon, I., & Suk Ryou, J. (2019). Ceramic processing research review on the self-healing concrete-approach and evaluation techniques. *Journal of Ceramic Processing Research, 20*(1), 1–18.

Jaivignesh, B., & Sofi, A. (2017). Study on mechanical properties of concrete using plastic waste as an aggregate. *IOP Conference Series: Earth and Environmental Science, 80*(1). https://doi.org/10.1088/1755-1315/80/1/012016

Jena, S., Basa, B., & Chandra Panda, K. (2020). A review on the bacterial concrete properties. *IOP Conference Series: Materials Science and Engineering, 970*(1). https://doi.org/10.1088/1757-899X/970/1/012004

Jonkers, H. M., Thijssen, A., Muyzer, G., Copuroglu, O., & Schlangen, E. (2010). Application of bacteria as self-healing agent for the development of sustainable concrete. *Ecological Engineering, 36*(2), 230–235. https://doi.org/10.1016/j.ecoleng.2008.12.036

JPNN.com. (2021). *Industri Baja Turut Berperan dalam Penurunan Emisi Gas Rumah Kaca.* www.jpnn.com/news/industri-baja-turut-berperan-dalam-penurunan-emisi-gas-rumah-kaca

Karatas, I. (2008). *Microbiological Improvement of the Physical Properties of Soils.* Arizona State University.

Khajuria, A., & Sharma, P. (2019). Use of plastic aggregates in concrete. *International Journal of Innovative Technology and Exploring Engineering, 9*(1), 4406–4412. https://doi.org/10.35940/ijitee.A5088.119119

Kim, O. S., Cho, Y. J., Lee, K., Yoon, S. H., Kim, M., Na, H., Park, S. C., Jeon, Y. S., Lee, J. H., Yi, H., Won, S., & Chun, J. (2012). Introducing EzTaxon-e: a prokaryotic 16s rRNA gene sequence database with phylotypes that represent uncultured species. *International*

Journal of Systematic and Evolutionary Microbiology, *62*(PART 3), 716–721. https://doi. org/10.1099/ijs.0.038075-0

Kim, S. B., Yi, N. H., Kim, H. Y., Kim, J. H. J., & Song, Y. C. (2010). Material and structural performance evaluation of recycled PET fiber reinforced concrete. *Cement and Concrete Composites*, *32*(3), 232–240. https://doi.org/10.1016/j.cemconcomp. 2009.11.002

Kwon, S., Nishiwaki, T., Kikuta, T., & Mihashi, H. (2013). Experimental study on self-healing capability of cracked ultra-high-performance hybrid-fiber-reinforced cementitious composites. *Sustainable Construction Materials and Technologies*. www.researchgate.net/ publication/268152482_Enter_titleExperimental_Study_on_Self-Healing_Capability_ of_Cracked_Ultra-High-Performance_Hybrid-Fiber-Reinforced_Cementitious_Com posites

Li, J., Wu, Z., Shi, C., Yuan, Q., & Zhang, Z. (2020). Durability of ultra-high performance concrete—a review. *Construction and Building Materials*, *255*, 119296. https://doi.org/ 10.1016/j.conbuildmat.2020.119296

Li, P., Wang, X., Su, M., Zou, X., Duan, L., & Zhang, H. (2021). Characteristics of plastic pollution in the environment: a review. *Bulletin of Environmental Contamination and Toxicology*, *107*(4), 577–584. https://doi.org/10.1007/s00128-020-02820-1

Liu, C., Xu, X., Lv, Z., & Xing, L. (2020). Self-healing of concrete cracks by immobilizing microorganisms in recycled aggregate. *Journal of Advanced Concrete Technology*, *18*(4), 168–178. https://doi.org/10.3151/jact.18.168

Marzouk, O. Y., Dheilly, R. M., & Queneudec, M. (2005). Reuse of plastic waste in cementitious concrete composites. *Proceedings of the International Conference on Cement Combinations for Durable Concrete*, *June 2014*, 817–824. https://doi.org/10.1680/ ccfdc.34013.0089

Megasari, S. W., Yanti, G., & Zainuri, Z. (2016). Karakteristik Beton Dengan Penambahan Limbah Serat Nylon Dan Polimer Concrete. *SIKLUS: Jurnal Teknik Sipil*, *2*(1), 24–33.

Ming, Y., Chen, P., Li, L., Gan, G., & Pan, G. (2021). A comprehensive review on the utilization of recycled waste fibers in cement-based composites. *Materials*, *14*(13), 3643. https://doi. org/10.3390/ma14133643

Monishaa, M., & Nishanthi, S. (2008). *Experimental Study on Strength of Self-Healing Concrete*. Universiti Malaysia Pahang, November.

Muslikin, B. T., & Masagala, A. A. (2020). *Penggunaan Bakteri Untuk Regenerasi Beton Yang Retak "Self Healing Concrete" Dengan Metode Pengujian Kuat Lentur* [Universitas Teknologi Yogyakarta]. http://eprints.uty.ac.id/5806/

Nasim, M., Dewangan, U. K., & Deo, S. V. (2020). Effect of crystalline admixture, fly ash, and PVA fiber on self-healing capacity of concrete. *Materials Today: Proceedings*, *32*, 844–849. https://doi.org/10.1016/j.matpr.2020.04.062

Nishiwaki, T., Koda, M., Yamada, M., Mihashi, H., & Kikuta, T. (2012). Experimental study on self-healing capability of FRCC using different types of synthetic fibers. *Journal of Advanced Concrete Technology*, *10*(6), 195–206. https://doi.org/10.3151/jact.10.195

Nkwachukwu, O., Chima, C., Ikenna, A., & Albert, L. (2013). Focus on potential environmental issues on plastic world towards a sustainable plastic recycling in developing countries. *International Journal of Industrial Chemistry*, *4*(1), 34. https://doi.org/10.1186/ 2228-5547-4-34

Nugroho, A., Satyarno, I., & Subyakto, S. (2015). Bacteria as self-healing agent in mortar cracks. *Journal of Engineering and Technological Sciences*, *47*(3), 279–295. https://doi. org/10.5614/j.eng.technol.sci.2015.47.3.4

Parker, L. (2019). *The World's Plastic Pollution Crisis Explained*. nationalgeographic.com. www.nationalgeographic.com/environment/article/plastic-pollution

Paul, S. C., van Zijl, G. P. A. G., & Šavija, B. (2020). Effect of fibers on durability of concrete: a practical review. *Materials*, *13*(20), 1–26. https://doi.org/10.3390/ma13204562

Pinto, P., & Brasileiro, F. (2021). Self-healing concrete: background, development, and market prospects. *Biointerface Research in Applied Chemistry*, *11*(6), 14709–14725. https://doi.org/10.33263/BRIAC116.1470914725

Qureshi, T., & Al-Tabbaa, A. (2020). Self-healing concrete and cementitious materials. In N. Tasaltin, P. S. Nnamchi, & S. Saud (Eds.), *Advanced Functional Materials*. IntechOpen.. https://doi.org/10.5772/intechopen.92349

Ragaert, K., Delva, L., & Van Geem, K. (2017). Mechanical and chemical recycling of solid plastic waste. *Waste Management*, *69*, 24–58. https://doi.org/10.1016/j.wasman.2017.07.044

Rohden, A. B., Camilo, J. R., Amaral, R. C., Garcez, E. O., & Garcez, M. R. (2020). Effects of plastic waste on the heat-induced spalling performance and mechanical properties of high strength concrete. *Materials*, *13*(15), 3262. https://doi.org/10.3390/MA13153262

Romano-Armada, N., Yañez-yazlle, M. F., Irazusta, V. P., Rajal, V. B., & Moraga, N. B. (2020). Potential of bioremediation and PGP traits in streptomyces as strategies for bio-reclamation of salt-affected soils for agriculture. *Pathogens*, *9*(2), 117. https://doi.org/10.3390/pathogens9020117

Safiuddin, M., Kaish, A. B. M. A., Woon, C. O., & Raman, S. N. (2018). Early-age cracking in concrete: causes, consequences, remedial measures, and recommendations. *Applied Sciences (Switzerland)*, *8*(10), 1730. https://doi.org/10.3390/app8101730

Saifee, S. N., Lad, D. M., & Juremalani, J. R. (2015). Critical appraisal on Bacterial Concrete. *IJRDO—Journal of Mechanical and Civil Engineering*, *1*(3 SE-Articles), 10–14. www.ijrdo.org/index.php/mce/article/view/532

Schlangen, E., & Sangadji, S. (2013). Addressing infrastructure durability and sustainability by self healing mechanisms—recent advances in self healing concrete and asphalt. *Procedia Engineering*, *54*, 39–57. https://doi.org/10.1016/j.proeng.2013.03.005

Shah, K. W., & Huseien, G. F. (2020). Biomimetic self-healing cementitious construction materials for smart buildings. *Biomimetics*, *5*(4), 1–22. https://doi.org/10.3390/biomimetics5040047

Sidiq, A., Gravina, R. J., Setunge, S., & Giustozzi, F. (2019). Microstructural analysis of healing efficiency in highly durable concrete. *Construction and Building Materials*, *215*, 969–983. https://doi.org/10.1016/j.conbuildmat.2019.04.233

Silva, F. B., Boon, N., De Belie, N., & Verstraete, W. (2015). Industrial application of biological self-healing concrete: challenges and economical feasibility. *Journal of Commercial Biotechnology*, *21*(1), 31–38. https://doi.org/10.5912/jcb662

Simon, K., & Milad, M. (2019). *Recycling Plastic Bottle into Synthetic Fiber Developing and Designing a Model for Recycling Plastic Bottle into Synthetic Fiber View project Smart Car Parking System View project* (Nomor December). International University of Business and Agriculture Technology.

Snoeck, D. (2015). *Self-healing and Microstructure of Cementitious Materials with Microfibres and Superabsorbent Polymers* [Ghent University]. https://biblio.ugent.be/publication/7010896

Stanaszek-Tomal, E. (2020). Bacterial concrete as a sustainable building material? *Sustainability (Switzerland)*, *12*(2), 696. https://doi.org/10.3390/su12020696

Subhashini, S., Yaswanth, K., & Prasad, D. S. V. (2018). Study on strength and durability characteristics of hybrid fibre reinforced self-healing concrete. *International Journal of Engineering & Technology*, *7*(4.2), 21. https://doi.org/10.14419/ijet.v7i4.2.19993

Van Tittelboom, K., & De Belie, N. (2013). Self-healing in cementitious materials—a review. *Materials*, *6*, Nomor 6. https://doi.org/10.3390/ma6062182

Vekariya, M. S., & Pitroda, P. J. (2013). Bacterial concrete: new era for construction industry. *International Journal of Engineering Trends and Technology (IJETT)*, *4*(9), 4128–4137. www.ijettjournal.org/archive/ijett-v4i9p181

Vijay, K., & Murmu, M. (2019). Self-repairing of concrete cracks by using bacteria and basalt fiber. *SN Applied Sciences*, *1*(11), 1–10. https://doi.org/10.1007/s42452-019-1404-5

Wiktor, V., & Jonkers, H. M. (2011). Quantification of crack-healing in novel bacteria-based self-healing concrete. *Cement and Concrete Composites*, *33*(7), 763–770. https://doi.org/10.1016/j.cemconcomp.2011.03.012

Xu, H., Lian, J., Gao, M., Fu, D., & Yan, Y. (2019). Self-healing concrete using rubber particles to. *Materials*, *12*, 1–16.

Yang, S., Yue, X., Liu, X., & Tong, Y. (2015). Properties of self-compacting lightweight concrete containing recycled plastic particles. *Construction and Building Materials*, *84*, 444–453. https://doi.org/10.1016/j.conbuildmat.2015.03.038

Yin, S., Tuladhar, R., Shi, F., Shanks, R. A., Combe, M., & Collister, T. (2015). Mechanical reprocessing of polyolefin waste: a review. *Polymer Engineering and Science*, *55*(12), 2899–2909. https://doi.org/10.1002/pen.24182

Zander, N. E., Sweetser, D., Cole, D. P., & Gillan, M. (2015). Formation of nanofibers from pure and mixed waste streams using electrospinning. *Industrial and Engineering Chemistry Research*, *54*(37), 9057–9063. https://doi.org/10.1021/acs.iecr.5b02279

Zhu, B., Wang, D., & Wei, N. (2022). Enzyme discovery and engineering for sustainable plastic recycling. *Trends in Biotechnology*, *40*(1), 22–37. https://doi.org/10.1016/j.tibtech.2021.02.008

Zulkernain, N. H., Gani, P., Chuck Chuan, N., & Uvarajan, T. (2021). Utilisation of plastic waste as aggregate in construction materials: a review. *Construction and Building Materials*, *296*, 123669. https://doi.org/10.1016/j.conbuildmat.2021.123669

8 Graded Concrete

Towards Eco-friendly Construction by Material Optimisation

M. Mirza Abdillah Pratama,[1] Poppy Puspitasari,[2]
and Hakas Prayuda[3]*
[1] Faculty of Engineering, Universitas Negeri Malang,
Indonesia
[2] Faculty of Engineering, State University of Malang,
Indonesia; Center of Advanced Materials and Renewable
Energy, State University of Malang, Indonesia
[3] Department of Civil Engineering, Faculty of Engineering,
Universitas Muhammadiyah Yogyakarta, Indonesia
[*] Corresponding author: M. Mirza Abdillah Pratama, mirza.
abdillah.ft@um.ac.id

CONTENTS

DOI: 10.1201/9781003320746-8

8.1   INTRODUCTION

8.1.1   Environmental Issue from Construction Sector

The industrial sector is responsible for 6% of all carbon dioxide (CO_2) emissions in the atmosphere (Sanjuán et al., 2020). According to another source, the cement manufacturing process emits 8% of CO_2, which can result in greenhouse gas impacts, which in turn can lead to significantly higher global temperatures and a changing climate (Flower & Sanjayan, 2007). This has anything to do with the activity of clinkering limestone ($CaCO_3$) and clay as raw materials for cement (calcination). Limestone is calcined at high temperatures in a cement furnace to make lime (CaO), which results in a 50% CO_2 waste discharge. The burning of fossil fuels in clinker accounts for up to 40% of all emissions. The fuel used to mine and transport raw materials accounts for the final 10% of emissions. According to data, CO_2 emissions from this process account for half of all cement produced. Until 2018, 4,500 million metric tons of cement were manufactured, allowing the quantity of emissions polluting the environment and their consequences to be determined (Cruz Juarez & Finnegan, 2021). Carbon capture and storage, the development of innovative cement, clinker replacement, the use of alternative non-fossil fuels in the calcination process, and energy efficiency have all been attempted to reduce CO_2 emissions (Imbabi et al., 2012).

8.1.2   Concrete as Preferable Material

Until now, concrete is still often used in various types of construction because of its advantages and is related to the energy used during the production process, implementation costs and economic aspects, the need for easily obtained materials, the reliability of its mechanical characteristics, and flexibility in the production process. Concrete production is energy efficient. Concrete is the only building material that produces the least amount of carbon dioxide. When compared to other materials, such as steel, the quantity of energy consumed in the manufacture of concrete is low. Cement, for example, requires only 450 to 750 kWh/ton of energy, whereas reinforced cement requires 800 to 3,200 kWh/ton of energy. Steel manufacture, for example, might require up to 8,000 kWh/ton of energy. Concrete is also recyclable and can be used as a roadbed and parking lot subbase. As a result, it is the best solution for decreasing carbon footprint and other building waste (Cantini et al., 2021).

Concrete is cheap and effective. Concrete is one of the most efficient and cost-effective building materials available. Concrete materials are of low maintenance as compared to other architectural materials, which means that they save money on maintenance and rebuilding. Concrete is incredibly affordable since the essential materials – cement, water, and aggregates – are all inexpensive and abundantly available around the world. Furthermore, unlike steel and wood, concrete structures do not need to be coated or painted on a regular basis. Coating and painting can be done on a regular basis, lowering maintenance expenses in comparison to other structures (Akadiri et al., 2012).

Concrete is a local product. Concrete is a common item since it can be found all around the world. When constructing buildings, suppliers do not need to carry materials long distances because they can manufacture from any location using local resources. As a result, shipping and pollution are reduced, while the local economy benefits (Torres et al., 2016).

Concrete is strong and long-lasting. Concrete is built to survive a long time, with a lifespan that is double that of other building materials (Ghali et al., 2016). It also gets stronger with time, lowering total ownership costs and the environmental impact of reconstruction. As a result, because it contributes to the compressive strength of reinforced concrete structures, it is the favoured alternative for carrying big loads. Concrete is also resistant to rust and rot, making it extremely long-lasting (Eid & Saleh, 2021). Its architectural structures are long-lasting and have water resistance features. Concrete constructions have proven to be robust in the aftermath of extreme flooding events (Cerè et al., 2017). Concrete is utilised to build structures in or near waterways, such as dams and waterfront projects, because it is appropriate for underwater and submerged applications.

Concrete is a fire-resistant material. Concrete is more resistant to high temperatures than wood or steel. It has a low heat conductivity and can resist heat for up to 6 hours (Callejas et al., 2017; Kodur, 2014). As a result, the material can be used in high-temperature and blast applications. Calcium silicate hydrate is the principal binder of concrete, and it can endure temperatures up to 910°C. It has the ability to store a significant quantity of heat from the environment, giving enough time for rescue in the event of a fire.

Concrete has the ability to be flexible. Concrete has a special flexibility that allows architects to create a wide range of buildings. It's a malleable building material that can be moulded into any shape, texture, or form. At a construction site, fresh concrete can be poured in a variety of configurations to create the desired shapes and sizes (Kirschke & Sietko, 2021). Concrete's fluidity makes it a favourite among designers and architects, despite the fact that it is strong and functional when hardened. By altering the mix, it may be cast into intricate shapes and formations. Innovations such as ultra-high-performance concrete and photocatalytic concrete have been made possible by technological advancements (D'Alessandro et al., 2017).

Furthermore, concrete can take on a variety of additions, looks, surface textures, and shapes. Every year, new types of concrete are developed, including mixtures with high-performance properties for specific uses. Concrete may be utilised in a variety of applications and can be customised to match even the most unusual requirements. It also does not require any special conditions to solidify; therefore, it can harden at room temperature. Concrete's moisture and high-temperature resistance make it an excellent candidate for environments that require these properties, such as undersea construction and high-heat manufacturing (Kirchhof et al., 2020).

8.1.3 Effort on Material Optimisation

Although concrete has been the most common building material for decades, it has a number of negative effects on the environment (Hashemi et al., 2015). Cement is the most important component in concrete production; however, it produces a certain amount of greenhouse gas emissions. It accounts for roughly 8% of global CO_2 emissions. CO_2 is released in two ways during the process. First, it necessitates the expenditure of energy to heat a kiln to high temperatures. Second, another transition occurs during the mixing of cement into concrete: calcination. When limestone is heated, it transforms into quicklime. On the other hand, the global population is increasing. This means more structures. Cement production continues to increase year after year, resulting in increased CO_2 emissions. Cement manufacture necessitates a significant quantity of energy. Smaller elements, such as sand and stone, are sourced from natural resources and environments. It also results in a large demand for fresh water, accounting for roughly 20% of global industrial water usage (Boretti & Rosa, 2019). These considerations demonstrate why identifying actual alternatives must help mitigate some, if not all, of these disadvantages.

Concrete's environmental credentials are being improved by innovators. Self-healing concrete, concrete printing, pollution-absorbing cement, waste material-incorporated concrete, and other innovations have already been developed. Self-healing concrete is a well-established technology that incorporates a bacteria-based healing agent to combat the unavoidable problem of fracture formation in concrete constructions (Vermeer et al., 2021). This technology is now limited by the high expense of producing this restorative ingredient. On other concrete technology, one of the most difficult aspects of three-dimensional (3D) printing concrete is coming up with a practical way for defining, establishing, and testing the required material properties of fresh and hardening concrete (Zhang et al., 2019). Designing the rheological parameters of printable concrete is a difficult task that frequently entails conflicting requirements arising from different stages of the manufacturing process. In green concrete development, the resulting compressive strength might vary depending on the green materials used (Al-Mansour et al., 2019). The rate of water absorption is great. In comparison to standard concrete, it has high shrinkage and creep qualities. Green concrete has a lower flexural strength. Green concrete constructions have a shorter lifespan than regular concrete structures. To understand the final concrete qualities, a full life cycle analysis of green concrete with many parameters is required.

In fact, applying lean construction concepts at all stages of building is one of the simplest ways to cut CO_2 emissions (Jang et al., 2022). The quantity of trash generated by project activities can be minimised to a bare minimum by applying lean construction. Concrete can be developed optimally to avoid overproduction, which wastes money and pollutes the environment.

8.2   CONCRETE AS STRUCTURAL MATERIALS

8.2.1   Understanding Concrete Behaviour

Concrete is well known for its strength (Wilkie & Dyer, 2021). Regardless, there are a few defects in concrete structural sections that must be considered when designing them. Concrete is made up of three fundamental components: water, aggregates, and Portland cement. When Portland cement comes into contact with water, it chemically reacts to form a rigid matrix that keeps the particles together. As a result, the material has relatively homogeneous strength in all directions. Concrete can withstand a great deal of pressure. When subjected to tensile loads, however, its resistance is considerably diminished, and disintegration is frequently abrupt (Surahyo, 2019).

Reinforcement, typically in the form of steel, fibreglass, or carbon fibre bars, is used to improve the tensile strength and ductility (ability to stretch and bend prior to failure) of concrete parts. Reinforcement, often known as rebar, has a much higher tensile strength than concrete. The use of rebar not only enables the fabrication of smaller and lighter sections, but it also aids in crack management (Tiberti et al., 2017). We must keep in mind that, whether reinforced or not, cracks in concrete elements will form naturally once in service. Concrete cracks can be quite disturbing to users of the structure because they are often imperceptible to the naked eye and sometimes quite large. To evaluate the importance of fractures, structural engineers must examine the structural behaviour of concrete materials. These can be stretched in four different ways: compression, tension, flexure, and shear.

8.2.2   Basic Mechanism in Reinforced Concrete Beam

When compared to other materials such as steel and wood, reinforced concrete is the dominant material utilised in the world of building construction. Over the past three decades, concrete technology has advanced rapidly. In this modern day, the demand for quality materials that are high in efficiency, have a long service life, and are resistant to weather fluctuations has become the primary goal for the development of concrete technology (Adaloudis & Bonnin Roca, 2021; Gebremariam et al., 2020; Gettu et al., 2018; Khan et al., 2021; Moriconi, 2007). Reinforced concrete is used in a variety of constructions, including residential buildings, bridges, stadiums, road pavements, dams, retaining walls, tunnels, bridges across valleys (viaducts), drainage and irrigation systems, water tanks, and so on. Concrete, mechanically, is a material that is strong in compression but weak in tensile strength (Krasnikovs et al., 2007). As a result, concrete can crack if the load it bears generates tensile stress greater than its tensile strength. Because of the stress mechanism, steel reinforcement is used in concrete to make a reinforced concrete structure. Moments caused by external loads create compressive and tensile forces in a concrete beam construction, resulting in compressive and tensile stresses in the cross section. Because concrete is very weak at resisting tension, the cross section of the concrete beam can collapse quickly in the tensile zone. Steel reinforcement (reinforcing bars) is inserted into concrete to produce a composite structure between the concrete and steel reinforcement, allowing the tensile force required to withstand moments in the fractured portion to be developed in the steel reinforcement. According to the description above, steel reinforcement is added to the cross section of the concrete that has the potential to suffer tension when holding the load to overcome the weakness of concrete in resisting tension. Steel reinforcement is highly beneficial in resisting tension because it has a very high tensile stress and tensile strain and it is ductile.

Bernoulli's law applies to bending structural members when the strain distribution along the cross-sectional height is considered to be linear (Szeptyński, 2020). Because the compressive stress–strain relationship of concrete is non-linear, and the presence of steel reinforcement in the cross section serves to transfer tensile forces when cracks occur in the cross section, the theory of flexural stresses in beams is not used in the design of reinforced concrete beams. In general, the behaviour of reinforced concrete beams in resisting bending is classified into five stages (Li et al.,

2020). Where the crack has not yet formed, the value of the strain caused by the working moment is quite tiny; hence, the stress distribution obtained is still essentially linear. The relationship between moment and curvature in the cross section is also linear in this case. Cracks will form near the lower edge of the portion experiencing maximum moment if the operating load is continuously increased. When the tensile tension at the bottom edge exceeds the tensile strength of the concrete, cracks form. When tensile force is applied to the concrete at the fracture position, it is transferred to the steel reinforcement, reducing the effective concrete cross section in resisting the moment. The beam stiffness is similarly lowered at this point, but the stress distribution remains close to linear. If the load is increased further, the steel reinforcement will finally fail. Following yielding, the beam curvature rapidly increases with a modest increase in moment.

Based on the aforementioned description of the behaviour of the beam cross section, it is considered that the foundation for flexural concrete theory is as follows (Rusch, 1960): before bending, the cross section perpendicular to the axis of bending is a flat plane; after bending, it remains a flat plane; no slip exists between the concrete and the steel reinforcement (at the same level, the strain in the concrete is the same as the strain in the steel); using the stress–strain relationship of concrete and steel, stress in concrete and reinforcement may be computed from strain; the tensile strength of the concrete is ignored when calculating the flexural strength of the section; when the compressive strain of concrete approaches the compressive limit strain, it is supposed to collapse; and the stress–strain relationship of concrete can be assumed to be square, trapezoidal, or parabolic, among other possibilities.

8.2.3 DISCOVERING RESEARCH GAPS FOR OPTIMISATION

Steel reinforced concrete members bend their load-carrying capability due to the integration of concrete compression and steel tensile strength. Concrete tensile capacity is overlooked by codes since it is insignificant in comparison to compressive strength. In theory, providing a low concrete strength to the layers in tension saves money and protects the environment.

8.3 CHARACTERISTICS OF GRADED CONCRETE

8.3.1 HISTORY OF GRADED CONCRETE

A functionally graded material (FGM) is defined as a material with properties that gradually change within the microstructure of components, such as modulus of elasticity, density, and Poisson's ratio. This material was the first to propose the use of FGM as an alternate material (Cavanagh et al., 2012). The FGM project was later developed for spacecraft in Japan in 1984. The idea behind this study was to mix metal and ceramic materials in a progressive transition, eliminating stress discontinuities and reducing difficulties connected with delamination, which were common in laminate structures. This concept also aimed to develop a new material that was resistant to high temperatures. FGM could provide material with optimal performance at a cheaper cost. FGM offers the advantage of boosting the residual stress distribution, improving thermal resistance and crack resistance, and lowering the stress intensity factor, thereby making it an excellent candidate for further development (Roesler et al., 2007; 2008).

FGM microstructures were classified into three types: continuously graded microstructures, discretely graded microstructures, and graded multi-phase microstructures (Rhee, 2007). In terms of microstructure, the composite material, which includes graded concrete, is categorised as a heterogeneous material. Because the model contains an excessive degree of freedom, the analysis at the microstructural level takes a long time to complete. To address this issue, the model is idealised into a structure composed of a number of layers with uniform material properties, a process known as layer-wise homogenisation. The structure's degree of freedom (DOF) will be reduced by applying this method, and the analysis will be faster. To simplify the model into homogenous FGM, layer-wise homogenisation was utilised (Miedzińska, 2017). When the composite material was described as particles scattered on a structure, FGM was modelled as a heterogeneous material. Composite materials were thought to be linearly distributed with regard to the specimen's height. The non-linear analysis application was used to examine both types of models until the ultimate load was attained. According to the results, some significant layers in the homogeneous model created a stress–strain curve that was closely convergent with the heterogeneous models.

FGM on concrete elements: beam, plate, and shell have been explored in various works. Research was conducted on corrugated sheets with functionally graded fibre cement. These conclusions were drawn from phenomena in which, during bending tests, stress tended to concentrate at the top and bottom fibres of its plate, whereas stress at the neutral axis was imperceptible. As a result, the decision to distribute the fibre evenly throughout the structure becomes less effective. Fibre should be dispersed progressively throughout its height, collecting at the plate's bottom, to lower production costs without sacrificing structural capacity, particularly in reaction to the modulus of rupture (Dias et al., 2010). Other researchers substituted fibre to improve beam element ductility, increase flexural strength, and optimise load distribution (Mastali et al., 2014; Shen et al., 2008). They used compression techniques during casting to guarantee the fibre was dispersed gradually, and the beam was studied using a scanning electron microscope (SEM). According to another study, X-ray computed tomography should be used to determine the distribution of fibre on fibre-reinforced concrete (FRC).

A simpler FGM can be created by combining two concrete mixtures with vastly differing properties in a moulded structural part. To achieve a seamless quality transition, intense compaction procedures must be used to avoid the formation of a cold-joint, which produces stress discontinuities when loading. Compaction can be done mechanically or vibratorily. According to previous research, printing graded concrete can be accomplished by first laying a high-strength concrete combination at the bottom and then placing a low-strength concrete mixture on top (Han et al., 2016). This is done so that the coarse aggregate, which is mostly present in the low-strength concrete mixture, can decrease by gravity quickly after laying, resulting in a natural quality gradation. For further action, the cast element can be reversed, resulting in a layer of high-strength concrete on the top fibre and a layer of low-strength concrete on the bottom fibre.

8.3.2 COMPRESSION STRENGTH OF GRADED CONCRETE

Pioneer research on graded concrete was performed through experimental study on grading pattern modelling in concrete by combining two types of concrete mixtures with considerable compressive strength differences (Gan et al., 2015). The strength of the material was measured by the lowest concrete quality in the graded concrete specimens that were moulded using a unique implementation approach. Other researcher investigated the impact of graded concrete on concrete strength by combining design characteristics of 20 MPa and 60 MPa to produce graded concrete specimens. According to experimental results and finite element analysis, the compressive strength of graded concrete is governed by the strength of the weakest material, but the stiffness of graded concrete is affected by the strength of the strongest material (Pratama, 2015). Investigation was done on the effect of two compressive strengths of concrete on graded concrete and the results revealed that the compressive strength of concrete was influenced by a weak layer of concrete, but its stiffness was influenced by a combination of low-strength and high-strength concrete strength (Figure 8.1). The strain gauge reading on the surface of the graded concrete specimen indicates that the strain is greatest at the bottom fibre. Lower quality materials with lower rigidity will experience greater displacement for the same load level. The strain response has a greater gradient in the top layer, indicating a stronger stiffness as compared to the bottom layer, due to the predominance of failure of the material at the bottom (Han et al., 2016).

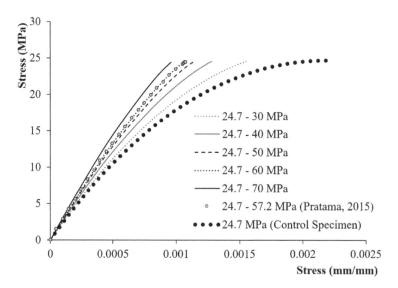

FIGURE 8.1   Stress–strain relationship of graded concrete and reference specimen (Pratama et al., 2019a).

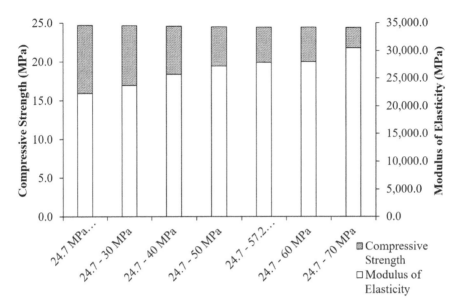

FIGURE 8.2 Compressive strength and modulus of elasticity of graded concrete and reference specimen (Pratama et al., 2019a).

8.3.3 MODULUS OF ELASTICITY OF GRADED CONCRETE

The modulus of elasticity is defined as the slope of the stress–strain curve in the linear elastic zone at approximately 30% of peak load. A high modulus of elasticity implies the ability to endure big stresses under low-strain conditions, which means that the concrete can withstand significant stresses caused by loads that occur at a low strain. The modulus of elasticity of concrete is affected by the age of the concrete, the aggregate and cement characteristics, the speed of loading, and the kind and size of the specimen. The modulus of elasticity is a value that indicates the stiffness and resistance of concrete to deformation by analysing changes in the height of the concrete cylinder as a result of loading. The modulus of elasticity of concrete is governed by the concrete's stress–strain relationship. The modulus of elasticity of graded concrete is intermediate between that of low-strength concrete and that of high-strength concrete (Figure 8.2).

8.4 FLEXURAL BEHAVIOUR OF GRADED CONCRETE BEAMS

8.4.1 STRUCTURAL PERFORMANCE AND SERVICEABILITY REQUIREMENT

The structure must be able to safely handle the design loads without causing excessive stress to the material and must have deformations that are within the allowed range (Basteskår et al., 2019). The ability of a structure to carry loads without experiencing excessive stress is attained by including the consideration of safety into the design of structural parts. The stress level in the structure can be determined at a level that is judged still acceptable safely, and such that excess stress in the material, for example, evidenced by the existence of fractures, does not occur, by choosing the size and shape of the structural parts and the materials employed. Essentially, this is the strength criterion, and it is a critical foundation. Another component of a structure's serviceability is whether or not the deformation generated by the load is within acceptable limits (Gilbert, 2011). Excessive deformation can result in excessive stress in a structure member. Furthermore, because extreme deformation is visible to the naked eye, it is frequently undesired. It should be highlighted that simply because a structure deforms considerably does not imply that it is unstable. Large deflections or deformations can be linked to dangerous constructions; however, this is not always the case. The stiffness of the structure governs the deformation. The stiffness of a structure is strongly reliant on the kind, size, and distribution of its materials. To achieve the requisite stiffness, more structural elements are frequently required than to meet the structural strength requirements. Movement in the structure is related to, but not the same as, deformation. The actual speed and acceleration of a structure carrying dynamic loads can be sensed by building users in numerous scenarios, causing discomfort. One example is movement in relation to a multi-story building subjected to wind loads. There are standards in place for this, such as speed and acceleration limitations. Control is done by structural stiffness and damping characteristics adjustments. A structure's serviceability is described as good performance under service load conditions, which can be represented in terms of deflection that is less than the limit requirements with non-structural elements such as walls, partitions, and ceilings remaining undamaged; control of cracks so that they are not noticeable, or to prevent water from entering and corroding steel or deteriorating concrete and requirements for service, such as vibration or noise.

8.4.2 DEVELOPMENT OF GRADED CONCRETE AS REINFORCED CONCRETE BEAMS

The usage of graded concrete can be used in the production of beam elements. Where the concrete material excels in compressive strength, higher-strength concrete is concentrated in the compression fibres, whereas lower-strength concrete is concentrated in the tensile fibres (Aylie et al., 2015). The design concept is based on the reinforced concrete structure design principle, in which tensile stress in structural elements is regulated by reinforcing steel. Because of the presence of compression-tensile conditions in the cross section of reinforced concrete beams, reinforced concrete beams can be constructed with a variety of concrete characteristics (quality grading). Because the strength of the concrete is negligible, the lower section of the beam subjected to tensile pressures does not require high concrete strength. In contrast to the top portion, which is dependent on the strength of the concrete, it necessitates the use of high-quality concrete. Because the strength of low-strength and high-strength concrete is combined, graded concrete offers stronger stiffness properties than uniform quality concrete (Pratama et al., 2021). The use of graded concrete in the construction of beam elements can lower the value of deflection, allowing the structure's service standard to be met.

The use of graded concrete to the elements of multi-storey building beams has no significant influence when compared to the use of high-quality concrete (Aylie et al.,

2015). Graded concrete beams require only 2.3 per cent additional reinforcement to meet the needs of the primary reinforcement, resulting in a 1.5 per cent greater deflection than high-grade reinforced concrete beams. Studies utilising finite elements have confirmed that putting graded concrete on beam elements can improve beam element performance, as evidenced by the improved moment–curve relationship of the beam compared to low-grade concrete beams (Pratama et al., 2019a; Pratama et al., 2019b). Furthermore, it is indicated that graded concrete beams have higher shear resistance than traditional reinforced concrete beams, as demonstrated by experimental research and finite element analysis with Abaqus (Pratama et al., 2019c). On the preceding research, graded concrete beams with three layers of quality, in which concrete with higher strength flanking lower strength, have significantly better structure behaviour than graded concrete beams with two layers of quality, with high quality on compression fibre and low quality in tensile fibres. This is due to the fact that laying high-quality concrete on tensile and compressive fibres can boost the strength of the concrete's adhesion to reinforcement while also enhancing the concrete's resistance to the compressive stress that occurs (Pratama et al., 2019d). Concrete beams have a higher crack resistance when placed in this manner, making the beams stronger to sustain the load that happens.

8.5 PROSPECT OF GRADED CONCRETE ON MULTI-STOREY BUILDING

Indonesia is one of the world's regions with geological faults and a high intensity of earthquakes. Most parts of Indonesia are susceptible to earthquake vibrations, including mild, moderate, and strong earthquakes. Therefore, designing the building structure with earthquake force in mind must be a priority and must be given special consideration. The zoning of structural elements that have the potential to experience plastic hinges in the design of earthquake-resistant structures, particularly high-rise buildings and bridge pillars, must be detailed in such a way that the structural elements have adequate ductility and are able to absorb vibration energy due to earthquakes so that the structure does not collapse. According to the Indonesian earthquake-resistant building standards, the design philosophy of earthquake-resistant building structures is that building structures that sustain strong earthquake loads may be damaged but must not collapse. According to this theory, the structure is allowed to deform inelastically (Pangestu & Pratama, 2021).

The configuration of the building is a significant consideration when selecting a building structural system. To the greatest extent practicable, the intended building's shape is roughly symmetrical, with the centre of mass and the centre of stiffness reasonably close together to avoid excessive eccentricities. Irregular shape is one of the factors that might lead to stress concentration in specific places. Furthermore, prevent structural imperfections in the horizontal and vertical directions as much as possible because one of them generates stress concentration. If structural irregularities, which may be created by architectural forms or space needs, cannot be avoided, certain standards must be addressed, such as precise features as needed by SNI-1726-2019 and SNI-2847-2019.

The magnitude of the probability that the load will be surpassed within a specific time, the level of ductility of the structure experiencing it, and the extra strength contained in the structure all contribute to the earthquake load value. If there is a significant earthquake, the internal forces that arise in the structural elements may surpass the calculated internal forces. The design earthquake is defined as an earthquake having a 2% chance of exceeding the magnitude during the structure's life of 50 years.

Previous study has shown that graded concrete can increase material stiffness, increasing the degree of service of the structure as evidenced by a decrease in deflection rate (Damayanthi et al., 2021). The enhanced stiffness is induced by the higher modulus elasticity of concrete, which is a combination of low- and high-quality concrete. Modulus elasticity of graded concrete is a type of contribution that comes from both poor- and high-quality concrete features. Investigations into the performance of structures that employ graded concrete as building materials need to be expanded using expansive metrics (Prayuda et al., 2021). The observation was carried out in a simulated program utilising Etabs. At Etabs, a three-story bay building is modelled in three dimensions.

Evaluation was carried out on base shear, fundamental period, storey stiffness, storey drift, and storey displacement. Base shear is a reaction that happens at the base of a building as a result of the load that works on the structure of the building, including the specified dead-, life-, and seismic load. The result option is used to generate the base shear reaction computationally. The larger the shear force created in buildings of uniform quality, the higher the quality of concrete applied to the building. This is due to the fact that the higher the quality of the concrete used, the more stiff the structure will be, so that the base of the building is rigid and provides tremendous restrictions. Because of these conditions, the building can elicit a stronger reaction and idealise a clamped pedestal that is stiffer than the joint pedestal. The base shear force created in buildings with graded concrete is between the buildings which use higher and lower concrete quality (Table 8.1).

The fundamental period is the amount of time it takes a structure to create a specific shape mode. Several factors determine the amplitude of the basic period, including the stiffness of the structure, its weight, its height, the orientation of the column geometry, the existence of walls on the structure, and the influence of the crack cross section. In this instance, multiple variables become bound variables; therefore, the rigidity of the structure and the weight of the structure must be investigated further. The stiffness and weight of a structure will grow linearly as the quality of the concrete used in its construction improves. The increased rigidity of the structure will have an effect on the building's shorter natural period. This is supported by an equation

TABLE 8.1
Base Shear Force

Concrete Strength (MPa)	X-Direction (kN)	Y-Direction (kN)	Significance (%)
30	300	300	
Graded concrete	313	313	+4.42
50	348	348	

TABLE 8.2
The Natural Period of Structures

Mode Shape	Fundamental Period (Second)		
	30	Graded Concrete	50
1	0.41	0.40	0.34
Significance (%)		−3.83	

TABLE 8.3
Storey Stiffness

Storey Level	Storey Stiffness (kN/mm)		
	U30	Graded Concrete	U50
3rd	60.92	67.22	94.18
Significance (%)		+10.34	
2nd	90.62	98.29	140.94
Significance (%)		+8.46	
1st	129.89	138.15	201.81
Significance (%)		+6.36	

stating that the magnitude of the natural period is always inversely proportionate to the structure's stiffness. The results of the analysis imply that buildings that employ graded concrete can provide a higher level of occupant comfort since the structure experiences a shake for a longer amount of time as the value of the vibration shaking period increases (Table 8.2). This effect will be felt strongly, especially during a big earthquake, and buildings made of low-quality concrete are at risk of surpassing its elastic phase as a result of large tremors.

Because of workmanship characteristics, floor plates and beams on reinforced concrete structures are a type of unity of monolith constructions. This is due to the fact that formwork and concrete casting work are done concurrently, and it is not permitted to interrupt the work technique until the two aspects are completed. The computation of the beam's capacity in the analytical process also considers the effective part of the plate as part of the element, so that the plate components can indirectly improve the beam's bending capacity. When the structure as a whole is analysed, the components that give stiffness to each floor of the building are not just beams and plates. The presence of column elements flanking the beams, as well as the compressive axial force that works and is funnelled by the column, can increase a building's floor stiffness. The stiffness of the floor will grow with a normal floor plan design from the ground level to the top, as the effect of restraint applied by the column on the plates and beams increases. This is according to the findings of the investigation, which reveal that the most basic floor has more rigidity than the upper floor (Table 8.3). The stiffness value of the floor decreases as the elevation of the floor analysed increases, as does the effect of constraint supplied by the column. The use of the graded concrete principle to beam components can raise the value of floor stiffness. This is because the stiffness produced by the beam increases with the modulus elasticity of the material used in the beam components. This, of course, raises the restrictions on the flanking plates, increasing the rigidity of the building floor.

The extent of the relative movement that happens on each floor to the floor below in a building is referred to as storey drift. The higher the position of the floor elevation to the ground level, the greater the value of story drift (Figure 8.3). The shape of the building being evaluated as if it applies as a cantilever beam clamped to the base ground causes the enlargement of the value of storey drift. The presence of an earthquake load that acts in the lateral direction of the building as though to give a beam shear force causes the structure to deviate. Gravitational loads on buildings, such as dead loads, additional dead loads, and live loads, act as axial forces on cantilever beams. Both earthquake loads and gravitationally acting loads create substantial variations at the intermediate level and small deviations at the lower and upper levels. The minimal deviation value at the top level is logically related to the lack of a working living load; however, the size of the deviation value in the intermediate level region is produced by the accumulation of the value of the gravitational load as well as the earthquake load. The effect of enlarging the value of storey drift is not very substantial as the quality of concrete used to the construction improves.

The rigidity of the beam bending to the column and the condition of the pedestal on the structure model influence story displacement in structures with a bound height. The use of high-quality concrete on the beam changes the structure's shape mode from flexible to shear-dominated manner. Under shear-dominated manner conditions, the column will generate a double curvature curve, causing the structure's shape mode to

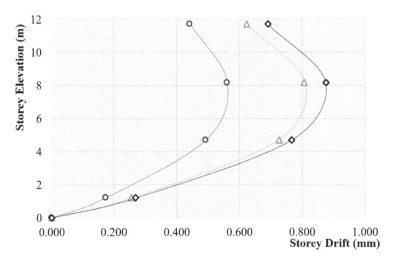

FIGURE 8.3   Storey drift.

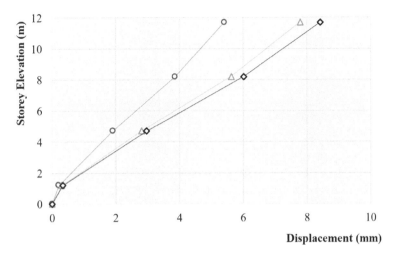

FIGURE 8.4   Story displacement.

collide with the curve, which tends to curve to form a convex pattern (Figure 8.4). The presence of footing at the base of the structure is represented by the use of this sort of pin pedestal at the bottom of the model. The behaviour of the construction structure as a whole is determined by the type of basic focus into flops. The emphasis of the flops prevents excessive floor variations on the lower floor, and the clamps also contribute to the restrictions in the column, making the column stiffer. To build a more stable construction, the structure will display a shear-dominated manner type shape mode from centre to bottom and a flexible type from centre to top.

8.6   CONCLUSION

Although concrete is regarded as a construction material with the lowest energy consumption when compared to other materials, its use with conventional designs and principles is deemed inefficient due to waste in the use of cement as an adhesive substance. This design is not only a waste of money, but also pollutes the environment because the cement manufacture and mobilisation process contributes to carbon gas emissions, which continue to rise exponentially due to development demands. Analytically, reinforced concrete structural elements do not necessitate the use of uniform concrete quality because the elements' tensile fibres have been reinforced with steel reinforcement, allowing the gradation of concrete quality to be used as an effort to optimise materials and realise lean construction. Graded concrete has the property of raising the modulus of elasticity of the material as well as the stiffness of the structure, allowing the serviceability level to be enhanced without incurring substantial costs. In terms of structural performance, the use of graded concrete can improve performance in terms of strength and user comfort, especially in earthquake-prone areas.

REFERENCES

Adaloudis, M., & Bonnin Roca, J. (2021). Sustainability tradeoffs in the adoption of 3D Concrete Printing in the construction industry. *Journal of Cleaner Production, 307*, 127201. https://doi.org/10.1016/j.jclepro.2021.127201

Akadiri, P. O., Chinyio, E. A., & Olomolaiye, P. O. (2012). Design of a sustainable building: a conceptual framework for implementing sustainability in the building sector. *Buildings, 2*(2), 126–152. https://doi.org/10.3390/BUILDINGS2020126

Al-Mansour, A., Chow, C. L., Feo, L., Penna, R., & Lau, D. (2019). Green concrete: by-products utilization and advanced approaches. *Sustainability, 11*(19), 5145. https://doi.org/10.3390/SU11195145

Aylie, H., Gan, B. S., As'ad, S., & Pratama, M. M. A. (2015). Parametric study of the load carrying capacity of functionally graded concrete of flexural members. *International Journal of Engineering and Technology Innovation, 5*(4), 233–241.

Basteskår, M., Engen, M., Kanstad, T., & Fosså, K. T. (2019). A review of literature and code requirements for the crack width limitations for design of concrete structures in serviceability limit states. *Structural Concrete, 20*(2), 678–688. https://doi.org/10.1002/suco.201800183

Boretti, A., & Rosa, L. (2019). Reassessing the projections of the World Water Development Report. *Npj Clean Water, 2*(1), 1–6. https://doi.org/10.1038/s41545-019-0039-9

Callejas, I. J. A., Durante, L. C., & de Oliveira, A. S. (2017). Thermal resistance and conductivity of recycled construction and demolition waste (RCDW) concrete blocks. *Revista Escola de Minas, 70*(2), 167–173. https://doi.org/10.1590/0370-44672015700048

Cantini, A., Leoni, L., De Carlo, F., Salvio, M., Martini, C., & Martini, F. (2021). Technological energy efficiency improvements in cement industries. *Sustainability, 13*(7), 3810. https://doi.org/10.3390/SU13073810

Cavanagh, J. R., Cross, K. R., Newman, R. L., & Spicer, W. C. (2012). Graded thermal barrier—a new approach for turbine engine cooling. *Journal of Aircraft, 9*(11), 795–797. https://doi.org/10.2514/3.59077

Cerè, G., Rezgui, Y., & Zhao, W. (2017). Critical review of existing built environment resilience frameworks: directions for future research. *International Journal of Disaster Risk Reduction, 25*, 173–189. https://doi.org/10.1016/J.IJDRR.2017.09.018

Cruz Juarez, R. I., & Finnegan, S. (2021). The environmental impact of cement production in Europe: a holistic review of existing EPDs. *Cleaner Environmental Systems, 3*, 100053. https://doi.org/10.1016/J.CESYS.2021.100053

D'Alessandro, A., Fabiani, C., Pisello, A. L., Ubertini, F., Luigi Materazzi, A., & Cotana, F. (2017). Innovative concretes for low-carbon constructions: a review. *International Journal of Low-Carbon Technologies, 12*(3), 289–309. https://doi.org/10.1093/IJLCT/CTW013

Damayanthi, W., Pratama, M. M. A., Susanto, P. B., & Prayuda, H. (2021). Stiffness of graded concrete beams on different cross-sectional dimension ratio and steel reinforcement ratio. *AIP Conference Proceedings, 2447*(1), 030014. https://doi.org/10.1063/5.0072560

Dias, C. M. R., Savastano, H., & John, V. M. (2010). Exploring the potential of functionally graded materials concept for the development of fiber cement. *Construction and Building Materials, 24*(2), 140–146. https://doi.org/10.1016/J.CONBUILDMAT.2008.01.017

Eid, M. S., & Saleh, H. M. (2021). Characterizations of cement and modern sustainable concrete incorporating different waste additives. In *Sustainability of Concrete with Synthetic and Recycled Aggregates*. IntechOpen. https://doi.org/10.5772/INTECHOPEN.100447

Flower, D. J. M., & Sanjayan, J. G. (2007). Green house gas emissions due to concrete manufacture. *The International Journal of Life Cycle Assessment*, *12*(5), 282–288. https://doi.org/10.1065/LCA2007.05.327

Gan, B. S., Aylie, H., & Pratama, M. M. A. (2015). The behavior of graded concrete, an experimental study. *Procedia Engineering*, *125*, 885–891. https://doi.org/10.1016/j.proeng.2015.11.076

Gebremariam, A. T., Di Maio, F., Vahidi, A., & Rem, P. (2020). Innovative technologies for recycling End-of-Life concrete waste in the built environment. *Resources, Conservation and Recycling*, *163*, 104911. https://doi.org/10.1016/j.resconrec.2020.104911

Gettu, R., Pillai, R. G., Santhanam, M., & Dhanya, B. S. (2018). *Ways of Improving the Sustainability of Concrete Technology Through the Effective Use of Admixtures*. Indian Institute of Technology Madras, Chennai, April 21.

Ghali, A., Gayed, R. B., & Kroman, J. (2016). Sustainability of concrete infrastructures. *Journal of Bridge Engineering*, *21*(7), 04016033. https://doi.org/10.1061/(ASCE) BE.1943-5592.0000862

Gilbert, R. I. (2011). The serviceability limit states in reinforced concrete design. *Procedia Engineering*, *14*, 385–395. https://doi.org/10.1016/j.proeng.2011.07.048

Han, A., Gan, B. S., & Pratama, M. M. A. (2016). Effects of graded concrete on compressive strengths. *International Journal of Technology*, *7*(5), 732. https://doi.org/10.14716/ijtech.v7i5.3449

Hashemi, A., Cruickshank, H., & Cheshmehzangi, A. (2015). Environmental impacts and embodied energy of construction methods and materials in low-income tropical housing. *Sustainability*, *7*(6), 7866–7883. https://doi.org/10.3390/SU7067866

Imbabi, M. S., Carrigan, C., & McKenna, S. (2012). Trends and developments in green cement and concrete technology. *International Journal of Sustainable Built Environment*, *1*(2), 194–216. https://doi.org/10.1016/J.IJSBE.2013.05.001

Jang, H., Ahn, Y., Roh, S., Yan, D., Jang, H., Ahn, Y., & Roh, S. (2022). Comparison of the embodied carbon emissions and direct construction costs for modular and conventional residential buildings in South Korea. *Buildings*, *12*(1), 51. https://doi.org/10.3390/BUILDINGS12010051

Khan, S. A., Koç, M., & Al-Ghamdi, S. G. (2021). Sustainability assessment, potentials and challenges of 3D printed concrete structures: a systematic review for built environmental applications. *Journal of Cleaner Production*, *303*, 127027. https://doi.org/10.1016/j.jclepro.2021.127027

Kirchhof, L. D., Lima, R. C. A. De, Neto, A. B. da S. S., Quispe, A. C., & Filho, L. C. P. da S. (2020). Effect of moisture content on the behavior of high strength concrete at high temperatures. *Revista Materia*, *25*(1), 12573. https://doi.org/10.1590/S1517-707620200001.0898

Kirschke, P., & Sietko, D. (2021). The function and potential of innovative reinforced concrete prefabrication technologies in achieving residential construction goals in Germany and Poland. *Buildings*, *11*(11), 533. https://doi.org/10.3390/BUILDINGS11110533

Kodur, V. (2014). Properties of concrete at elevated temperatures. *ISRN Civil Engineering*, *2014*, 1–15. https://doi.org/10.1155/2014/468510

Krasnikovs, A., Lapsa, V., & Eiduks, M. (2007). Non-traditional reinforcement for concrete composites—state of the art (Netradicionālais betona stiegrojums – mūsdienu problēmas stāvoklis). *Transport and Engineering (Mašīnzinātne Un Transports Mechanics)*, *24*, 191–200.

Li, Q., Guo, W., Liu, C., Kuang, Y., & Geng, H. (2020). Experimental and theoretical studies on flexural performance of stainless steel reinforced concrete beams. *Advances in Civil Engineering*, *2020*, 1–13. https://doi.org/10.1155/2020/4048750

Mastali, M., Mastali, M., Abdollahnejad, Z., Ghasemi Naghibdehi, M., & Sharbatdar, M. K. (2014). Numerical evaluations of functionally graded RC slabs. *Chinese Journal of Engineering, 2014*, 1–20. https://doi.org/10.1155/2014/768956

Miedzińska, D. (2017). New method of numerical homogenization of functionally graded materials. *Procedia Structural Integrity, 5*, 484–491. https://doi.org/10.1016/j.prostr.2017.07.148

Moriconi, G. (2007). Recyclable materials in concrete technology: sustainability and durability. *Proceedings of Special Sessions of First International Conference on Sustainable Construction Materials and Technologies*, 11–13.

Pangestu, S. F., & Pratama, M. M. A. (2021). Evaluasi Kinerja Struktur Gedung Bertingkat Menggunakan Pendekatan Desain Berbasis Kinerja (Studi Kasus: Gedung Pendidikan Rangka Beton Bertulang 7 Lantai). *Cantilever: Jurnal Penelitian Dan Kajian Bidang Teknik Sipil, 10*(2), 91–100. https://doi.org/10.35139/CANTILEVER.V10I2.110

Pratama, M. Mirza Abdillah. (2015). *An Experimental Study and Finite Element Approach to the Behavior of Graded Concrete* [Universitas Diponegoro]. https://doi.org/10.13140/RG.2.1.4082.4568

Pratama, M. Mirza Abdillah, Sthenly Gan, B., Han Ay Lie, H. A. L., & Nur Rahma Putra, A. B. (2019a). A numerical analysis of the modulus of elasticity of the graded concrete. *Proceedings of the 2nd International Conference on Vocational Education and Training (ICOVET 2018)*. https://doi.org/10.2991/icovet-18.2019.29

Pratama, M. Mirza Abdillah, Suhud, R. K., Puspitasari, P., Kusuma, F. I., & Putra, A. B. N. R. (2019b). Finite element analysis of the bending moment-curvature of the double-layered graded concrete beam. *IOP Conference Series: Materials Science and Engineering, 494*(1), 012064. https://doi.org/10.1088/1757-899X/494/1/012064

Pratama, M. M. A., Arifanda, W., Karyadi, K., Nindyawati, N., Sulaksitaningrum, R., & Prayuda, H. (2019c). Numerical and experimental investigation on the shear resistance of functionally graded concrete (FGC) beams. *IOP Conference Series: Materials Science and Engineering, 669*, 012055. https://doi.org/10.1088/1757-899X/669/1/012055

Pratama, M. M. A., Umniati, B. S., Nindyawati, N., Margaretha, D. I., Suhud, R. K., Puspitasari, P., & Permanasari, A. A. (2021). Effect of concrete strength configuration on the structural behaviour of reinforced graded concrete beams. *Journal of Physics: Conference Series, 1808*(1), 012017. https://doi.org/10.1088/1742-6596/1808/1/012017

Pratama, M. M. A., Vertian, T., Umniati, B. S., & Yoh, W. H. (2019d). Flexural behaviour of the functionally graded concrete beams using two-layers and three-layers configuration. *IOP Conference Series: Materials Science and Engineering, 669*, 012054. https://doi.org/10.1088/1757-899X/669/1/012054

Prayuda, H., Pratama, M. M. A., Novitasari, Y., Damayanthi, W., Arifurrizal, S., Puspitasari, P., & Permanasari, A. A. (2021). Seismic performance of low-rise building using graded concrete as flexural elements. *IOP Conference Series: Materials Science and Engineering, 1144*(1), 012014. https://doi.org/10.1088/1757-899X/1144/1/012014

Rhee, R. S. (2007). *Multi-scale Modeling of Functionally Graded Materials (FGMs) Using Finite Element Methods*. University of Southern California.

Roesler, J., Bordelon, A., Gaedicke, C., Park, K., & Paulino, G. (2008). Fracture behavior and properties of functionally graded fiber-reinforced concrete. *AIP Conference Proceedings, 973*, 513–518. https://doi.org/10.1063/1.2896831

Roesler, J., Paulino, G., Gaedicke, C., Bordelon, A., & Park, K. (2007). Fracture behavior of functionally graded concrete materials for rigid pavements. *Transportation Research Record: Journal of the Transportation Research Board, 2037*, 40–49. https://doi.org/10.3141/2037-04

Rusch, H. (1960). Researches toward a general flexural theory for structural concrete. *Journal Proceedings*, *57*(7), 1–28. https://doi.org/10.14359/8009

Sanjuán, M. Á., Andrade, C., Mora, P., & Zaragoza, A. (2020). Carbon dioxide uptake by cement-based materials: a Spanish case study. *Applied Sciences*, *10*(1), 339. https://doi.org/10.3390/APP10010339

Shen, B., Hubler, M., Paulino, G. H., & Struble, L. J. (2008). Functionally-graded fiber-reinforced cement composite: processing, microstructure, and properties. *Cement and Concrete Composites*, *30*(8), 663–673. https://doi.org/10.1016/J.CEMCONCOMP.2008.02.002

Surahyo, A. (2019). Physical properties of concrete. In: Concrete Construction. Springer, Cham. https://doi.org/10.1007/978-3-030-10510-5_3

Szeptyński, P. (2020). Comparison and experimental verification of simplified one-dimensional linear elastic models of multilayer sandwich beams. *Composite Structures*, *241*, 112088. https://doi.org/10.1016/J.COMPSTRUCT.2020.112088

Tiberti, G., Trabucchi, I., AlHamaydeh, M., Minelli, F., & Plizzari, G. (2017). Crack control in concrete members reinforced by conventional rebars and steel fibers. *IOP Conference Series: Materials Science and Engineering*, *246*(1), 012008. https://doi.org/10.1088/1757-899X/246/1/012008

Torre, A., Burkhart, A., Torres, A., & Burkhart, A. (2016). Developing sustainable high strength concrete mixtures using local materials and recycled concrete. *Materials Sciences and Applications*, *7*(2), 128–137. https://doi.org/10.4236/MSA.2016.72013

Vermeer, C. M., Rossi, E., Tamis, J., Jonkers, H. M., & Kleerebezem, R. (2021). From waste to self-healing concrete: a proof-of-concept of a new application for polyhydroxyalkanoate. *Resources, Conservation and Recycling*, *164*, 105206. https://doi.org/10.1016/J.RESCONREC.2020.105206

Wilkie, S., & Dyer, T. (2021). Design and durability of early 20th century concrete bridges in Scotland: a review of historic test data.*International Journal of Architectural Heritage*, 1–21. https://doi.org/10.1080/15583058.2020.1870776

Zhang, J., Wang, J., Dong, S., Yu, X., & Han, B. (2019). A review of the current progress and application of 3D printed concrete. *Composites Part A: Applied Science and Manufacturing*, *125*, 105533. https://doi.org/10.1016/J.COMPOSITESA.2019.105533

9 Performance of Surgical Blades from Biocompatible Bulk Metallic Glasses and Metallic Glass Thin Films for Sustainable Medical Devices Improvement

Yanuar Rohmat Aji Pradana,* Aminnudin, and Heru Suryanto
Faculty of Engineering, Universitas Negeri Malang, Indonesia
* Corresponding author: yanuar.rohmat.ft@um.ac.id

CONTENTS

DOI: 10.1201/9781003320746-9

9.1 INTRODUCTION: BACKGROUND AND DRIVING FORCES

Lately, the demand of technological and quality development in medical devices is escalated rapidly in order to increase the living standard, in terms of improving the health quality and life longevity of human being. Extensive medical devices are made from metallic alloys, with their metallurgical improvements leading to enhance the biocompatibility and biomechanical performance (Lin et al., 2013). As is well known, metallic implant was firstly utilized as dental application in early 200 CE (Whitesides & Wong, 2006) and the application was spreading in large scope of medical implants and surgical devices until now. Magnesium (Mg) and its alloys, zinc (Zn) and its alloys, and calcium (Ca) and its alloys are commonly developed for biodegradable materials, especially for implants aiming to temporarily support and to be used in fixation devices of bones such as screws and plates, and cardiovascular stents for many clinical cases, without secondary surgery for taking out these implants after their roles are no longer needed (Li et al., 2014). On the other hand, stainless steel (SS), carbon steel (CS), cobalt-chromium (Co-Cr) alloys, zirconium (Zr) alloys and titanium (Ti) alloys are widely applied as surgical devices, stents, artificial knee joints, hip joints substitution, plates for bone, and dental implants.

In medical fields, surgery operations are applied for clinical cases to make an opening in soft tissue (i.e., skin and muscle) for investigating a pathological condition, such as a disease or injury. The aims of surgery are developed into improving the bodily function, appearance, and repair a ruptured area considered unwanted. In terms of body appearance enhancement, skin grafting is widely implemented to reconstruct and fix the damaged skin from extensive trauma such as burns and wounds (Chu et al., 2019). For these applications, the surgical and skin-grafting blades must fulfill several criteria: high sharpness, comparable hardness, sufficient strength, good durability, and high biocompatibility (Chu et al., 2014, 2019; Krejcie et al., 2012). For these reasons, diamond blade is the best candidate for material choice among surgeons for precision surgery, such as ophthalmic surgeries, due to its extreme hardness and achievable sharpness in single crystal diamond structure (Charles H. Williamson, 2007). However, besides its material cost, the manufacturing of cutting edge geometries takes the larger portion for making this diamond-made surgical blade remain expensive. This issue comes out also for other precious metal alloy applications, such as Pt- and Pd-based alloys (Liu et al., 2011). Although the costly material and manufacturing issues of diamond and precious metal blade fabrication are expelled in most crystalline metals such as martensitic SS and CS, the polycrystalline structure of such materials having average grain size >10 μm and the regions of grain boundary would be scrapped easily during grinding and sharpening processes, resulting in irregular and wavy edge-tip of blade in microscale view (Jang et al., 2015; Liao et al., 2020; Tsai et al., 2012). As a consequence, the martensitic SS and CS are larger in edge radii than diamond blades having an average radii of ~300 nm and restrict them for reaching the sufficient sharpness of desired surgical blades having edge radii ranged from 20 to 50 nm (Krejcie et al., 2012). When the blade having such edge-tip surface problem is utilized to cut the soft tissue, the wound tissue would consequently generate fragmented surface by irregular tearing and snatching; therefore, it can induce trauma, pain, and slow recovery (Tsai et al., 2012, 2014). On the other hand, the wavy and high surface roughness of edge-tip serve the crack initiation sites that leads to localized fracturing under further loading during blade application that can be emphasized as materials' drawbacks in terms low strength and wear resistance, particularly for long incision operation reaching more than 30 cm (Chu et al., 2019). Theoretically, the problem of such crystalline metal materials could be overcome by the use of oxide glass and ceramics materials; however, limited toughness and ductility are still the crucial issues (Meagher et al., 2016). Furthermore, it is not economically feasible and environmentally friendly to utilize the disposable SS blades which lose their sharpness quickly and consequently need the application of multiple devices in performing single surgical incision.

Metallic glasses (MGs) or called amorphous alloys, in the form of its bulk and thin films, have recently attracted significant recognition from fundamental science and engineering application point of view over past 30 years due to their unique and superior properties characterized by the short-range, random atomic order; therefore, they exhibit no limitation found in conventional crystalline metal structure such as dislocation, grain boundary, etc. These specific random atomic structure could offer high strength and hardness, low Young's modulus, high corrosion resistance, high wear resistance, excellent biocompatibility, and sufficient antibacterial properties (Huang et al., 2009; Li & Zheng, 2016; Li et al., 2018). MGs soften above a glass transition temperature (T_g) of alloy before crystallization (T_x). In this supercooled liquid region (SCLR), MG becomes viscous and easy to be formed using low applied forces. At this relatively low processing temperatures (~100 K) and lack of solidification shrinkage, MGs could be formed into net and complex shapes through a small-force processing techniques such as forming and extrusion and injection blow molding (Rajan & Arockiarajan, 2021). Initially, most of MGs were produced from metal elements that are considered costly, such as Au, Ag, Pt, and Pd, to improve their glass-forming ability (GFA) (Li & Zheng, 2016; Nishiyama et al., 2012). GFA of any alloy is defined as how easily the alloy system could be produced into amorphous structure by direct cooling from its liquid phase. Higher GFA value indicates a lower cooling rate required to form the glassy structure and a higher critical size of MG cast resulted from casting process (Meagher et al., 2016). Although having superior mechanical and chemical properties compared with their counterparts, MGs without the precious elements mentioned above have limited archievable critical size (in the range of microns), restricting their application in engineering and biomedical fields. These limitations were then solved by the development of bulk metallic glasses (BMGs) with various alloy systems, namely, Zr-, Fe-, Ti-, Mg-, Co-, under a relatively lower critical cooling rates (less than 100 K/s) (Gong et al., 2016; Meagher et al., 2016) by implementing basic empirical rules proposed by Inoue (Akihisa Inoue, 2000): composed at least three components/elements, considerable atomic size difference between them (>12%), and negative heat of mixing between (major) the constituent elements. By this range of BMGs' critical size, it is highly possible to fabricate them into macroscale profile such as surgical blade/scalpel with the homogeneous, isentropic, flat edge-tip morphologies, and excellent sharpness due to the annihilation of grain boundaries.

In parallel, due to the brittleness of some BMG compositions such as Fe-based BMG, the BMG was utilized as sputtering target on the commercial blade substrate to form metallic glass thin film (MGTF) coating in order to enhance the MG alloy flexibility and ductility by reducing them into small scale (~10 μm) (Chu et al., 2019; Rajan & Arockiarajan, 2021; Tsai et al., 2014). Furthermore, the MGTF application is intended for using the MG more efficiently compared with the bulk one. Both BMGs and MGTF could increase significantly the hardness, sharpness, and durability of blades, reducing the friction between blade surfaces and soft tissue, and prevent the bacteria transmission. Blades with better sharpness and durability are able to provide a clean, long, and neat incision requiring less cutting forces, less pain for the patient, and less bacterial transmission, resulting in patient faster recovery (Chu et al., 2019). By considering the characteristics of BMG and MGTF materials, the excellence of those materials from several alloy systems, surgical blades, their fabrication techniques, insight of the cutting performance, and their future challenges in developing medical tools and other related biomedical parts are provided in this chapter.

9.2 POTENTIAL MG ALLOY SYSTEMS FOR BIOMEDICAL APPLICATIONS

BMGs have been popular for decades owing to a set of extraordinary properties that correspond to their randomly ordered atomic structure. Pol Duwez's Group from California Institute of Technology was the one that successfully reported the first ever metallic glass finding in the composition of $Au_{75}Si_{25}$ in 1960 (Klement Jun. et al., 1960). Since then, a set of MG alloy systems has been successfully developed by many researchers for large applications, including biomedical field. The comparison of mechanical properties of several biocompatible MGs, biocompatible metal alloys, and bioglass is presented in Figure 9.1. Indeed, different from implantable matters, the materials corrosion and ion release of several surgical devices such as surgical blades/scalpel are occurred only at minimum amount since the interaction with human body takes place for only in seconds during surgery operation. However, the toxic-containing composition, such as Ni, Cr, Be, and other harmful elements that are possibly released from, for instance, commercial crystalline Co-Cr alloy, SS, and Ti-Al-V alloy, must be excluded to avoid further reactions and infections when such elements reach a certain threshold. Therefore, further attempts have been made to develop much safer MGs composition with higher biocompatibility and better mechanical properties to fulfill the technological demand of biomedical applications. BMG alloys that have potential for non-absorbable and non-degradable medical devices use, such as blades, drills, and plates, are mainly based on Fe, Ti, and Zr with toxic and potentially harmful elements diminished from the alloy systems. Certain biocompatible compositions of BMG and MGTF are briefly discussed in the following sub-sections, covering their potential thermal, mechanical, and chemical, including biocompatibility and anti-microbial, properties.

9.2.1 FE-BASED MG

In comparison with Zr- and Ti-based MGs, Fe-based MGs are relatively less costly but still perform sufficient GFA to form large size, making the composition attractive

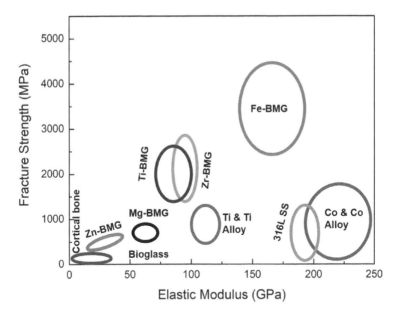

FIGURE 9.1 Mechanical properties comparison of several biocompatible MGs, biocompatible metal alloys, and bioglass (adopted from review works by Meagher et al., 2016; Yiu et al., 2020).

and economically feasible to be applied in large biomedical applications. This relatively good GFA enables the ability of MGs to be fabricated by conventional Cu-mold casting methods. Moreover, Fe-based MGs could perform extremely high fracture strength of more than 4000 MPa (Gu et al., 2007a), superior hardness of more than 1000 HV (kg/mm^2) (Gu et al., 2007b), and exceptional wear and corrosion resistance (Li & Zheng, 2016; Wang et al., 2009). Since the early development of Fe-based MGs in 1995, several alloy systems were successfully synthesized, such as in Fe-(Zr, Hf, Nb)-B and Fe-Co-Ln (lanthanide metal)-B (Inoue & Takeuchi, 2011); however, most of them are included as soft magnetic materials that are often needed in functional and structural applications such as power transformator and soft magnetic cores. Considering these magnetic properties, the systems are not suitable if applied in biomedical field; for instance, the presence of magnetic-performing implants in the body is inhibited for magnetic resonance imaging (MRI) diagnosis (F. M. Bui M. P. Mintchev, 2000; Shafiei et al., 2003). The magnetized implants might be magnetized by MRI instruments during the operation and produce image artifact, and therefore prevent the apparent diagnosis.

To overcome the problem described previously, non-magnetic Fe-based MGs were then developed in 2003 by Ponambalam's group, namely "amorphous steel" containing Fe, C, Cr, and Mo elements as similarly found in SS material (Ponnambalam et al., 2003). This amorphous steel could reach a critical size of 16 mm in diameter, and a variety of Fe-based MGs was further investigated. Previous study also demonstrated a higher MRI compatibility of Fe-based MG compared to that of 316L SS.

For thermal properties, MGTF with five different Fe content has been studied by Chen et al. where the higher Fe content clearly increased the crystallization temperature (T_x) to 989.2 K but lower SCLR (ΔT) of 26.3 °C (Chen et al., 2014). The larger SCL region (ΔT) was revealed in the $Fe_{41}Cr_{15}Mo_{14}C_{12}B_9Y_2Co_7$ system with 81 K (Tsai et al., 2014). For Fe-Cr-Mo-C-B-Y- MGs, both rod and thin film have been developed with the rapid rise of ΔT that was accomplished by the addition of a rare element in Fe-B-Nb system. Nd elements showed the larger ΔT and higher GFA for MGTF (Phan et al., 2011) caused by the substrate's strain-controlled annealing which occurred in MGTF. This elongation of ΔT clearly increases the opportunity of BMG superplastic forming for biomedical devices.

The hardness of Fe-based MGs is characterized by nanoindentation and microvickers for MGTF and BMG, respectively. For MGTF, the hardness was reported reaching 8.6 GPa in biocompatible $Fe_{37-33}Zr_{35}Nb_{21-26}$ system by Lou et al. (Lou et al., 2019). Moreover, $Zr_{38-29}Ti_{21-17}Fe_{37-49}$ MGTF could reach 9.3 GPa in hardness as reported by Chen et al. (Chen et al., 2014). For bulk one, the vickers' hardness reached 1122 HV in both $Fe_{63}Mo_{14}C_{15}B_6Er_2$ and $Fe_{55}Cr_8Mo_{14}C_{15}B_6Er_2$ (Gu et al., 2007b). Furthermore, in $Fe_{41}Cr_{15}Mo_{14}C_{12}B_9Y_2Co_7$ developed by Tsai et al., the hardness reached 1200 HV value (Tsai et al., 2014). With such level of hardness, the wear resistance is exceptional as well. Fe-based MGs also possessed the highest fracture strength than that on Zr- and Ti-based MGs reaching more than 4 GPa in $Fe_{49}Cr_{15}Mo_{14}C_{(13+x)}B_{(8-x)}Er_1$ (x = 2, 4, 5, and 6) and $Fe_{48}Cr_{15}Mo_{14}C_{15}B_6Er_2$ systems with relatively high plastic strain of 3.6% and the ductility could be significantly improved by modifying the alloy composition properly (Gu et al., 2007a, 2007b). The main shortcomings of the MGs alloy are the limited GFA and high elastic modulus compared to Ti- and Zr-based MGs; however, these limitations cause less problem in non-load-bearing implantable applications such as dental implants and surgical blades, because the critical size/thickness can reach at least 2 mm with such GFA (Meagher et al., 2016).

The chemical and environmental stability can be examined by the corrosion behavior of MGs study. The absence of grain boundaries in Fe-based MGs becomes a clear reason as to why the materials show high corrosion properties and are able to hinder the Ni ion release and localized attack in threatening biological environment after a long-term application as found in commercial 304 SS and 316L SS. This Ni ion release could cause several allergies and infections in human body. Considering the structure, Fe-based BMGs and MGTFs showed greater pitting potential and lower passivation current density in Hank's balanced salt solution (HBSS), artificial saliva, Ringer's solution, 5 wt% NaCl aqueous solutions, and artificial sweat solutions (Chen et al., 2014; Li et al., 2015; Lou et al., 2019; Ni et al., 2009; Obeydavi et al., 2020; Phan et al., 2011; Wang et al., 2009). By coating Fe-based MGTF on SS substrate, the MGTF exposes a sluggish rise of current density attributed by the passive layer and corrosion by-product layer formation. The corrosion behavior MG ribbons with the composition of $Fe_{51}Cr_{18}Mo_7B_{16}C_4Nb_4$ was analyzed in Ringer's solution showing a higher polarization resistance value than that on Ti-6Al-4V and 316L SS (Zohdi et al., 2011). Similarly, AISI420 SS was coated by both Fe-Zr-Nb and Fe-Zr-Ti MGTFs and the corrosion resistance could be enhanced up to 65 times higher (Chen et al., 2014; Lou et al.,

2019). The absence of any defects such as holes was clearly displayed in MGTF corroded surface compared with corrosion pits and by-products covered in that of uncoated SS. Tsai et al. reported by using HBSS the optimum alloy composition of $Fe_{41}Cr_{15}Mo_{14}C_{12}B_9Y_2Co_7$ BMG showing lower corrosion current density values and higher pitting potential values ($E_{pit} - E_{corr}$) of 1.356 V than those of 304 SS and 316 SS (Tsai et al., 2014). These higher corrosion resistances have a strong relationship with the random atomic arrangement, dislocation, and other crystalline defects absence, the involvement of alloying elements having high passivation capabilities such as Ti, Cr, Mo, and valve elements like Al, Zr, Nb, Hf, and Ta (Padhy et al., 2010). The other possible reason is the dense atomic packed structure due to the larger atomic size gap that possibly reduces free volume, vacancy, and void which can be the initial site of corrosion formation.

9.2.2 Ti-Based MG

Ti-based MGs become one of the preferable MG alloys for biomedical application owing to their relatively high mechanical properties, excellent biocompatibility, and low elastic modulus, as well as high resistance to wear and corrosion (Lin et al., 2013). The level of elastic modulus (55–110 GPa) is comparable to that of human bone (15–30 GPa), reducing the risk of stress-shielding effect caused by large modulus mismatches inducing insufficient and inhomogeneous loading between implant and its adjacent bones, and finally resulting in tissue loss and implant failure (Calin et al., 2013). Even though these MGs show high mechanical properties and are able to be cast into bulk shape, the Be or Ni elements considered toxic are required in the systems, limiting them for biomedical applications. Recently, the elimination of toxic element in Ti-based MG was attempted while maintaining their mechanical and corrosion properties to make them feasible to be applied in biomedical fields. Indeed, the low GFA of biocompatible Ti-based MG makes them difficult to form bulk size; however, attempts have been made to enable them prepared in larger/bulk size by making bulk metallic glass foams (BMG foams). By using vacuum hot pressing method, a set of $Ti_{42}Zr_{35}Ta_3Si_5Co_{12.5}Sn_{2.5}$ BMG foams with different porosity has been successfully fabricated by Nguyen et al. (Nguyen et al., 2019; Nguyen et al., 2020) and Wong et al. (Wong et al., 2022) from Jang's Group using different spacer materials of NaCl, Al, and Cu particles. These efforts enable the Ti system for both surgical and implantable devices applications.

The thermal stability and glassy nature of MGs are highly dependent on their thermal properties. Based on differential scanning calorimetry (DSC), when a constant increase of heat rate is applied into MGs heating, the stages occur as follows: (1) glass transition temperature (T_g), (2) crystallization temperature (T_x), (3) melting temperature (T_m), and (4) liquidus temperature (T_l). The length of region between T_g and T_x, or called supercooled liquid (SCL) region ($\Delta T = T_x - T_g$), becomes one of the parameters in determining the thermal stability of MG (Inoue). Both $Ti_{60}Nb_{15}Zr_{10}Si_{15}$ and $Ti_{40}Cu_{36}Pd_{14}Zr_{10}$ MGs showed relatively similar value of 49.3 K (Thanka Rajan et al., 2017, 2019). This value is far lower than those on $Ti_{42}Zr_{40}Ta_3Si_{15}$ BMG with 99 K and can be increased by the addition of Sn with 149 K on $Ti_{42}Zr_{42}Ta_3Si_{10}Sn_3$ BMG (Nguyen et al., 2020).

For biomedical applications, mechanical properties of Ti-based MGs play an important role in sustaining the several types of loading from surrounding tissues, bones, and blood vessels. As mentioned before, Ti-based BMG performed relatively low elastic modulus at around 78–115 GPa, high fracture strength (although lower than that on Fe-based BMGs) reaching 2500 MPa, plastic strain of 0.5%, and exceptional specific strength (Li & Zheng, 2016). However, most of Ni- and Be-free Ti-based BMGs with high GFA still contain Pd elements that are considered expensive, such as $Ti_{40}Zr_{10}Cu_{34}Pd_{14}Sn_2$ (Zhu et al., 2008); therefore, most of the lower cost biocompatible Ti-based MGs are in the form of ribbons. Considering this issue, nanoindentation technique becomes a reliable and convenient method to evaluate their mechanical properties such as elastic modulus and hardness. Both $Ti_{75}Zr_{10}Si_{15}$ and $Ti_{60}Nb_{15}Zr_{10}Si_{15}$ MG ribbons performed the compressive strength of more than 2000 MPa with the hardness of 790 and 660 HV, respectively. However, Pang et al. (Pang et al., 2015) successfully developed $Ti_{47}Cu_{38}Zr_{7.5}Fe_{2.5}Sn_2Si_1Ag_2$ BMG by arc melting and subsequent Cu-mold casting having a critical diameter of 7 mm. This material showed a comparable compressive fracture strength of 2080 MPa with considerable bulk plasticity. Furthermore, in the case of Ti-based MGs made from hot pressing of amorphous powders by Nguyen et al., $Ti_{42}Zr_{35}Si_5Ta_3Co_{12.5}Sn_{2.5}$ BMG established a compressive strength of 1342 MPa (Nguyen et al., 2020).

The corrosion rates of ternary, quartiary, or more components of MGs are strongly determined by the electrochemical nature of their main alloying elements; however, when the understanding is expanded into more detail, thermodynamic metastability, single-phase nature-dependent chemical homogeneity, and unusual atomic structure aspects are also decisive. Several simulated body fluids (SBFs) such as Ringer's solutions, HBSS, 1 mass% lactic acid, and phosphate-buffered solution (PBS) were often applied by researchers to study the corrosion resistance of Ti-based BMGs. Almost all studies revealed that Ti-based MGs exhibited passive behavior on the corrosion potential with reduced corrosion rate, localized corrosion susceptibility toward pitting corrosion, and localized corrosion resistance that is considered better than those on the crystalline counterparts such as 304 SS (Lin et al., 2014), 316L SS (Morrison et al., 2007), pure Ti, and Ti-6Al-4V (Calin et al., 2013; Oak et al., 2007; Pang et al., 2015). Ke et al. measured the electrochemical properties of Ti-based MGTF in SBF, where the corrosion resistance is higher by the higher content of Ti in the MGTF (Ke et al., 2014). When the MGTF-coated samples were immersed in electrolyte, passive layer was formed and the Ti element was oxidized to construct TiO_2 layer that can be a reason of the greater potential and corrosion resistance improvement (Rajan & Arockiarajan, 2021).

9.2.3   ZR-BASED MG

Zr-based MGs are the most attractive glassy alloys for many researchers and practitioners among other ternary and quaternary systems due to their remarkable GFA leading to extensive glass-forming range. This extremely high GFA was characterized by their atomic packing density where the glass stability is highly dependent on its atomic arrangement. As is well known, Zr atomic size of 0.155 nm is much larger and provides a sufficient size difference to other alloying elements to form BMGs with high GFA. This high GFA enables the alloy to fabricate into "bulk" form which is a common term for MGs having a section thickness or diameter of at least 1 mm. The thermal stability of Zr-based BMGs was represented in the large SCL region (ΔT), enlarging the potential of the bulk materials to be fabricated using superplastic forming by extensive applied temperatures and flow stresses.

Considering their high GFA, Zr-based BMGs are able to fabricate into large size by using direct Cu-mold casting. Nearly a 14-mm diameter BMGs rod was produced by Peker and Johnson in the composition of $Zr_{41.2}Ti_{13.8}Cu_{12.5}Ni_{10}Be_{22.5}$ or commonly called Vitreloy 1 (Peker & Johnson, 1993). In 1993, Inoue et al. successfully prepared Zr-Al-Ni-Cu BMGs with critical diameter of 16 mm and large SCL region (127 K) by water quenching and this system was successfully improved as $Zr_{55}Al_{10}Ni_5Cu_{30}$ BMGs with a critical diameter of 30 mm and a length of 50 mm by vacuum suction casting (Inoue et al., 1993; Akihisa Inoue & Zhang, 1996). However, such BMG alloys contain Be and Ni which are considered as toxic elements and are harmful for human cell after a certain time of exposure. In parallel, the high GFA of Ni- and Be-free Zr-based BMGs is also contributed by the involvement of expensive Pd element, which can be a limitation of biocompatible Zr-based BMGs development (Liu et al., 2011).

Recently, several Zr-based MGs were successfully synthesized using nontoxic, low Cu elements with exceptional mechanical properties in the form of bulk and thin film. Sun et al. successfully fabricated $Zr_{46}Cu_{37.6}Ag_{8.4}Al_8$ BMG using Cu-mold casting with the hardness, compressive fracture strength, and elastic modulus of 554 HV, 2158 MPa, and 92 GPa, respectively (Sun et al., 2014). The glassy alloy also showed relatively large plastic strain of 12%. Subsequently, based on research conducted by Wada et al., the absence of toxic elements in 180-mm-diameter $Zr_{56}Al_{16}Co_{28}$ BMG rod has no effect on its mechanical properties, where the alloys still performed a tensile strength of 1830 MPa and an elastic modulus of 83 GPa (Wada et al., 2009). Using the similar system of Zr-Al-Co, Li et al. tried to improve the mechanical properties by introducing $Zr_{54}Al_{17}Co_{29}$ monolithic BMG showing the fracture strength of 2220 MPa, and the strength was significantly enhanced to 3170 MPa by the addition of 1 at% of Ta particles (Li et al., 2018) with the higher level of plastic strain (8.6%). For thin film application, there are several compositions that were studied by many researchers having high hardness and wear resistance. $Zr_{48}Cu_{36}Ag_8Al_8$ MGTF showed great potential for biomedical application with hardness and elastic modulus values of 8.92 and 112.6 GPa, respectively, measured by nanoindentation (Thanka Rajan et al., 2016). Such modulus values are mostly lower than that on pure Ti (105 GPa), Ti-6Al-4V (110–125 GPa), and 316L SS (200 GPa). The high strength of Zr-based BMGs greatly enhances the opportunity of biomaterials production with thinner size, which provides a better deliverability in cardiovascular stents as well as greater holding power for fractured bones in bone screws (Li & Zheng, 2016). In other words, the smaller implant at the similar level of strength is much better for human body. Note that high elastic limit and low elastic modulus of this glassy alloy indicate the more uniformly external stresses distribution in human bones than crystalline materials, thus reducing stress concentration, avoiding stress shielding effects, and accelerating the healing process. The high hardness, good wear resistance, and better surface conditions (low friction coefficient) are also the advantages of Zr-based MGs for surgical device applications such as orthopedic drill (Li et al., 2016), surgical blades (Tsai et al., 2012), and coated needle (Bai et al., 2020; Chu et al., 2016).

Zr-based BMGs and MGTF coating show an enhanced corrosion resistance behavior compared to 316L SS, pure Zr, and Zr-based crystalline alloys, and pure Ti and Ti alloys such as Ti-6Al-4V in an extensive physiological-like media, such as HBSS (Huang et al., 2013; Ke et al., 2014; Li et al., 2018), Ringer's solution (Yiu et al., 2020), artificial saliva (Cai et al., 2017), PBS (Chang et al., 2015), HCl solution (Chuang et al., 2013), and NaCl saline solution (Chang et al., 2015; Morrison et al., 2004). The corrosion resistance behavior was indicated by a lower current density, higher pitting potential, and more protective layers formed on Zr-based MGs surface than those on the crystalline alloys. Besides the absence of crystal defects such as vacancy and grain boundary, the other reason of the high corrosion resistance of Zr-based MGs is related to the passive films formation, mainly composed by ZrO_2 and other Ti, Al, and Ag oxides (Li & Zheng, 2016; Rajan & Arockiarajan, 2021) on the alloy surface which enhances the resistance against pitting corrosion. Addition of small amount of Ag and/or Nb can also increase the stability from dissolution of ions and improves the corrosion resistance in different environments after a certain period of immersion (Hua et al., 2012; L. Liu et al., 2006).

9.3 BIOCOMPATIBILITY OF MGS

The biocompatibility of the materials chosen for biomedical applications is crucial. The word "biocompatibility" refers to a material's capability to interact with human body tissues without creating unacceptably high levels of harm. This ability is determined by a number of factors: (1) the human body's reaction to it, or the implant's cell biological activity, where the higher the bioactivity, the more biocompatible the implant is with surrounding tissues; and (2) material degradation in the human body environment as a result of insufficient wear and corrosion resistance. These phenomena could result in the release of the implant alloy system's constituent metal ions or particles into the body, generating a large amount of reactions, including allergic and toxic ones (Li & Zheng, 2016). Superior thermal, mechanical, and chemical properties of MGs described above give a clear argument for biocompatibility of these materials when compared with the current applied materials in medical field. For Fe-based MGs, there are several reports presenting the excellence of glassy alloys in cell viabilities. A quick grow of cell numbers of murine fibroblast cells (L929 cell and NIH3T3 cell) is observed during a culture period of 4 days. The cells with higher density are noticed forming an elongated and spindle shape. This high cell viability corresponds to the protective effect of resulted compact oxide film, anticipating the metal ion release in the biological fluid (Wang et al., 2012). Also, the Fe-Zr-Nb MGTF-coated specimen exposed exceptional optical density (OD) value for the 5th day, and the toxic-free coatings enabled a better cell growth and viability, demonstrating their elevated biocompatibility (Lou et al., 2019). The Fe and Nb ion contents are below the threshold considered less toxic, and a trace amount of Fe also exists in human blood as hemoglobin (Lou et al., 2019; Wang et al., 2012). Furthermore, earlier research has shown that Fe-based BMGs have a greater MRI compatibility than 316L SS (Fang et al., 2008), making it a suitable candidate for biomedical materials and therapeutic devices utilized in MRI bio-imaging and diagnostics.

Turning to Ti-based MGs, researchers have investigated the biocompatibility of these alloys using both *in vitro* cell response (via MTT/CCK8 assay and cell morphology observations) and *in vivo* through animal implants. The results have confirmed that Ti-based BMGs showed improved biocompatibility than the crystalline Ti-6Al-4V and Ti-45Ni alloys for the *in vitro* human osteoblast SaOS$_2$ cells (Oak et al., 2009). Several biocompatibility experiments were performed on these two BMGs, as well as the comparative samples of Ti-6Al-4V and pure Cu, in co-culture with the L929 murine fibroblast cell line. The Ti$_{40}$Zr$_{10}$Cu$_{36}$Pd$_{14}$ BMG and Ti-6A-l4V exhibit the best biocompatibility in terms of cellular adhesion, cytotoxicity, and metallic ion release affection, with cells remaining connected to the petri dish with comparable adhesion and exhibiting the spindle shape following direct contact testing. In addition, the Ti$_{40}$Zr$_{10}$Cu$_{36}$Pd$_{14}$ BMG had a very low Cu ion release level, which matched the MTT findings (Li et al., 2017). The antibacterial activity of Ti-based MGTF with Cu was investigated using several bacterial stains, including *Escherichia coli* (Subramanian, 2015). At reduced magnification, scanning electron microscope (SEM) revealed the reduction of bacterial colonies and cell density on the MGTF-coated material. In addition, Rajan et al. conducted *in vivo* tests on Albino rabbits using Ti-based MGTF extracts and found that there were no signs of erythema or edema in the investigations, and there were no signs of toxicity. The irritation index for erythema and edema was zero for the control, and 0.03 for all animals (rabbits) for the specimen extract, indicating nontoxicity (Thanka Rajan et al., 2019). On the other hand, hydroxyapatite (HA), which is considered as the main chemical composition in human natural bones and teeth, was found to have accumulated on the Ti-based BMG alloy surfaces after 15 days immersed in SBF (Li & Zheng, 2016), indicating that Ti-based BMGs are biocompatible.

Both *in vitro* cell response and *in vivo* animal implantation have been used to examine the biocompatibility of Zr-based MGs. *In vitro* cell culture (MTT/ CCK8 assay and cell morphology observations) revealed that Zr-based BMGs had superior biocompatibility than typical crystalline 316L SS, Zr, and Zr-based alloys, and Ti and Ti-based alloys including human umbilical vein endothelial cells (Li et al., 2013), MG63 human osteoblast-like cells (Li et al., 2012; Li et al., 2013), human aortic endothelial cells (HAECs) and human aortic smooth muscle cells (HASMCs), preosteoblast cells MC 3T3 E1 (Hua et al., 2012), murine fibroblast cells (L929 cell and NIH3T3 cell) (Wang et al., 2011), and *in vivo* rabbits implantations (Liu et al., 2009). Mouse fibroblast cells (L929 fibroblast cells) (Cai et al., 2017; Balasubramanian Subramanian et al., 2015) and preosteoblast MC3T3-E1 cells (Hua et al., 2014; Jabed et al., 2019) were used in *in vitro* cell line studies such as MTT assay and cell morphological evaluations. Different biological allergy pathways may be triggered by the substance implanted in the human environment, resulting in inflammation and implant loosening, among other consequences. The nontoxic nature of MGTF is confirmed by the MTT assay, which shows cell viability and proliferation. In Zr-Co-Al BMG system, even though Co ions are well known to be cytotoxic, the total mass of Co ions released by a unit surface area of $Zr_{56}Al_{16}Co_{28}$ BMGs (0.3 ng/mm^2) is significantly lower than the Co–Cr–Mo alloy (5.8 ng/mm^2) (Wada et al., 2009).

Another aspect of biocompatibility assessment lies in considering the antibacterial activity of the material. Antibacterial activity of Zr-based MGTF with Ag and Cu has been examined utilizing various bacterial strains such as *E. coli* and *S. aureus* (Chu et al., 2014). The bacterial colony reduction on the Zr-based MGTF surface and the uncoated surface was observed by SEM analysis at reduced magnification. The release of weakly bound Ag or Cu ions, or redox reaction and chemical ionization, is considered to be responsible for MGTFs' antibacterial properties (Etiemble et al., 2017). The electrostatic force of attraction attracts metal ions to the negatively charged bacterial cell walls, which disrupts the cell membranes, reacts with proteins, and affects the bacteria, eventually killing them (Chu et al., 2014; Balasubramanian Subramanian et al., 2015). In parallel, by antibacterial test, no or only a few bacterial colonies were found on the BMG sample, indicating that the surface is capable of eradicating bacteria at a 99.99 percent level. All of the findings suggested that $Zr_{58.6}Al_{15.4}Co_{18.2}Cu_{7.8}$ BMG might be used in surgical devices (Han et al., 2021).

9.4 BLADE FABRICATION TECHNIQUES

9.4.1 MACHINING THE BULK SHAPE OF MGS

To accommodate the superior characteristics of BMG that has been successfully developed by many researchers, an effort is needed to shape the material into a standard product that is ready to be used, one of which is through the machining process. Conventional machining was carried out on Zr-based BMGs by several researchers using high-speed machining, in which light emission occurs during cutting causing crystallization. This can be reduced by lowering the cutting speed and feed rate, and applying wet machining (Bakkal, 2009; Chen et al., 2017; Maroju et al., 2018). Li et al. (Jang's Group) developed an orthopedic drill from $Zr_{48}Cu_{35.3}Al_8Ag_8Si_{0.7}$ BMGC with Ta particle reinforcement through conventional machining followed by water-cooled polishing which resulted in better drilling ability with 73% lower thrust and drilling force fluctuations than the commercial orthopedic drills (Li et al., 2016).

In addition to using conventional machining, BMG cutting can also be done through nonconventional machining, such as laser machining and electrochemical machining. However, William and Lavery reported that in BMG processing involving heat energy such as laser cutting, crystallization is unavoidable so that it will degrade the characteristics of the material (Williams & Lavery, 2017). Meng et al. successfully fabricated the $Ni_{72}Cr_{19}Si_7B_2$ BMG complex product through wire electrochemical micro machining (WECMM), however, bubbles generated at narrow machining distances in the electrolyte will reduce cutting accuracy (Meng et al., 2017).

Therefore, machining using electrical discharge machining (EDM) can be an option in Zr-based BMG cutting because the role of the dielectric fluid other than as an insulator and an expulsion system is as a coolant so that the evaporation process of the material can be localized and crystallization can be inhibited. Tsai et al. (Jang Group) succeeded in fabricating $Zr_{48}Cu_{35.3}Al_8Ag_8Si_{0.7}$ BMG into a surgical blade formed from BMG plates using electrical discharge cutting (EDC) and sharpened through a polishing process which has better performance than commercial scalpels (Tsai et al., 2012).

Huang and Yan revealed the effect of voltage and capacitance on micro-EDM Zr-based BMG where low voltage and capacitance results in smaller craters and recast layers, lower surface roughness, and also lower material removal rate (MRR) (Huang & Yan, 2015, 2016). Crystallization in the form of ZrC phase on the surface and internal phase below the surface decreases with increasing voltage and decreasing capacitance. In addition, the crystallization that occurs in micro-EDM is much smaller when compared to wire-EDM. Yeo et al. reported that the use of low-input energy reduces surface roughness by 43–51% and burr width by 63% on micro-EDM $Zr_{57}Nb_5Cu_{15.4}Ni_{12.6}Al_{10}$ BMG, while at low-input energy, tube-shaped electrodes experience a smaller wear ratio when compared to rod-shaped electrode.

In the EDM process, the factors that determine process productivity, electrode wear, and surface quality are divided into electrical aspects (duration, polarity, current, voltage, and electrode distance), dielectric (oil type, flash point, viscosity, and flushing mechanism), electrode (material, shape, and size), and workpiece (material and size) (Patel & Maniya, 2018). Koyano et al. and Zhu et al. sequentially revealed the use of silicon and graphite electrodes to modify MRR, roughness, and tool wear in the cutting process (Koyano et al., 2019; Zhu et al., 2018). Wang et al. reported the use of a small electrode size to minimize the deviation of the machining yield size (Wang et al., 2019). In addition, current and polarity are the most decisive parameters on MRR, roughness, and tool wear in cutting using sinking-EDM (Gowthaman et al., 2018). Pradana et al. conducted an EDM cutting on $Zr_{54}Co_{29}Al_{17}$ BMG and found that in all levels of pulse on-time, the material removal rate (MSS), surface roughness, and surface hardness were enhanced (Pradana et al., 2020).

In addition, ultrasonic machining (USM) is also an alternative method for cutting high hardness and brittle materials such as BMG. Productivity and quality of the USM technique are highly dependent on the mechanical characteristics of the workpiece (hardness and fracture toughness), tool characteristics (hardness, impact strength, surface condition, and design), abrasive characteristics (hardness, roughness, size, distribution, and viscosity), and process settings (power input), static loading, amplitude and frequency of vibration (Kumar, 2013). The material removal mechanism in USM is initiated by micro-brittle fracture that occurs on the surface of the glass, followed by crack propagation and the intersection of the median and radial cracks caused by repeated impact loads of abrasive grains which eventually causes chipping resulting in material removal. The applied machining process has a significant effect on the production of better dimples so that it can increase profits in terms of increasing mechanical characteristics, improving surface quality, reducing tool wear, and increasing MRR, so that the overall cost of the product is lower when viewed from an economic point of view (Maroju & Jin, 2018). In the cutting of BMG using USM as described by Kuriakose et al., the temperature increase did not occur so that the amorphous structure was maintained. In addition, machining BMG via micro-USM results in high MRR with very low tool wear rates with minimal burrs. The quality of the microholes produced is very good due to the high feed rate and grit number of the abrasive slurry (Kuriakose et al., 2017).

9.4.2   Surgical Blade Fabrication from BMGs

As mentioned previously, commercial surgical blade (scalpel) and dermatome were recently made from martensitic SS and CS. The surgical blade consisted of numerous types of shape depends on the application needed. To shape BMG into surgical blade, several steps are needed. The common arc melting and Cu-mold suction casting were applied for processing most of the metallic glasses and BMG plates, which resulted in having 2 or 3 mm of thickness. After a systematical analysis of structure by X-ray diffraction (XRD) to ensure the glassy structure of alloy, wire-electrical discharge machining (WEDM) was used to form the shape surgical blade on BMG. The shape was made to be identical with the commercial one, including handle and cutting edge tip. WEDM was subsequently utilized to reduce the thickness of BMG. The more identical in thickness to commercial blade (± 0.2 mm), the more realistic and valid data will be recorded during the cutting test. WEDM is chosen for BMG processing method due to its feasibility to control the temperature using dielectric flushing to avoid extensive surface crystallization and high cutting precision characteristics (Huang & Yan, 2016). The blade-shaped BMG was then grinded, sharpened, and polished exactly at the cutting edge to form an angle of 30°. These processes were conducted on centrifugal sand paper machine with the gradual sand paper grit from smaller grit (#150, coarse) to larger grit (#5000, fine). To fine-polish and sharpen the cutting edge, Al_2O_3 micropowder was used to make the sharp, flat, and shiny edge tip. The cutting edge-tip of the blade was finally observed using SEM with high magnification to ensure the shape appropriateness and XRD to check if any structural change occurs during blade fabrication. These steps of BMG blade-making are considered the simplest way due to the manual works in polishing and sharpening.

9.4.3   Hybrid Process for Surgical Blade Manufacturing from BMGs

Due to the opinion of machining and superplastic forming drawbacks to produce >50 nm in edge radii, thermally assisted micro-drawing was discovered to be capable of producing extremely small features, such as the edge radii. As a result, it was determined to conduct further research by using a combination process of thermally assisted micro-molding and thermally-assisted micro-drawing to form bulk metallic glass within the SCLR, as proposed by Krejcie et al. (2012). To carry out the hybrid manufacturing process, a testbed that is capable of accurately molding the BMG sample within its SCL region and then precisely drawing the molded blade was created. A ball screw actuator controls the upper die. The positional gap between the upper die and lower one is measured with a resolution of 100 nm using a linear variable differential transformer (LVDT) that provides feedback to the process controller. The BMG sample is drawn using an additional actuator with sub-micron precision. This actuator includes a load cell and an adapter plate for mounting the BMG sample. Tool steel inserts are used in the critical geometries of both the upper and lower dies for maximum strength, higher precision, and ease of maintenance. The lower die is adjustable, allowing for precise alignment of the upper and lower dies during molding to create a homogeneous gap thickness. Heat is provided by resistive cartridge heaters located within the upper and lower dies equipped with thermocouple to provide accurate measurement of heating temperature.

Prior to the start of the micromolding process, the dies are heated to a predetermined temperature within the SCL region of BMG. The first two steps involve shaping the rake face and determining the initial geometry of the blade. The next step involves stopping the upper die when the remaining material thickness is around 20 μm. The last step provides the subsequent micro-drawing process, where the BMG sample is drawn, causing it to plastically deform until it fails along a sharp edge.

9.4.4   Magnetron Sputtering for MGTFs Deposition

The MGTF is usually deposited onto a substrate to enhance the surface properties, including mechanical, chemical, and thermal properties. In making MGTF-coated blade, the selected composition of BMG was firstly fabricated using common arc melting and suction Cu-mold casting from 99.9% purity of the respective constituents. The process was often intended to produce BMG plate. After a set of investigations including XRD analysis, thermal analysis by DSC, and compositional analysis using scanning electron microscopy–energy dispersive X-ray analysis (SEM-EDX), the BMG plate was then sectioned using WEDM, lathed, and assembled into a complete disc as a sputtering target. By magnetron sputtering process, the BMG disc was then deposited onto a substrate (usually a commercial blade) in an extreme vacuum environment by controlling the base pressure, working pressure, sputtering power, sputtering time, and gas flow. A thin layer of BMG was then coated on the substrate and subsequently analyzed using several film characterization methods for further cutting and scratch tests.

9.5   PERFORMANCE OF BMG AND MGTF-COATED BLADES

The performance of BMG and MGTF-coated blade is commonly evaluated by considering several aspects: blade-sharpness index (BSI), adhesion (in MGTF), and ductility of cutting edge-tip. Due to a limited loading and thermal change during skin and tissue incision, most of MGs structures are unchanged after cutting, and therefore crystallization can be neglected. Several studies have reported that all of BMG and MGTF-coated blades display a relatively smoother and more shiny surfaces visually compared with the commercial one. This was also confirmed by SEM observation showing a flat cutting edge-tip compared with MSS blade, which although performing flat at the edge, the near edge area is relatively wavy and inhomogeneous (Chu et al., 2019; Jang et al., 2015; Tsai et al., 2012).

9.5.1   Blade Sharpness Index

BSI is the parameter to evaluate the blade sharpness, which follows the fracture mechanics principle that the critical energy for initiating a crack in large amount of soft materials is obtained by a cutting tool with given geometry and mechanical properties. The lower force for initiating indentation on the substrate corresponds to the sharper blade, resulting in a better cut in terms of neat and smooth wound for quick

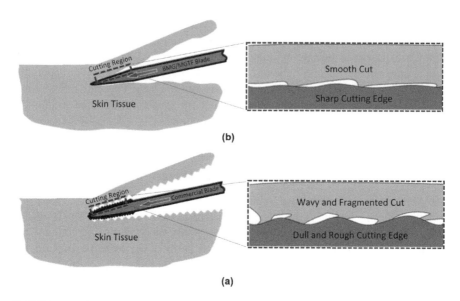

FIGURE 9.2   Schematic illustration of (a) commercial and (b) BMG/MGTF-coated blade in skin incision. Smooth and sharp cutting edge of BMG and MGTF-coated blades results in neat and smooth cutting (adopted and edited from Chu et al., 2014).

recovery, as shown in Figure 9.2. Initially, using a specially prepared rig for cutting test, the blade was clamped at the tool post upward the substrate clamped location. The substrate is usually made from rubber material and is considered as simulated skin tissue. The movement of upper part of the rig was controlled and equipped by a universal testing machine; therefore, the load and displacement were recorded. The cutting was then applied with a low quasi-static rate for two passes: the first pass was for determining the load-displacement curve (X) and the second pass was for obtaining the pure friction (P). The force only for initial cutting indentation (pure cutting force) (X-P) was then calculated. J_{IC} was firstly calculated to evaluate further BSI following Equation 9.1 as proposed by McCarthy et al. (2007, 2010):

$$J_{Ic} = \frac{(X-P)u}{dA} \tag{9.1}$$

where X is the force acting on the blade, P is the pure friction force working between the blade and the substrate, u is the blade vertical displacement, and dA is the increase of newly created surface area due to cutting equal to $2ht$, as shown in Figure 9.3. The interaction strength (X - P) between the blade and the substrate was measured at each step of the test. Then, the BSI can be determined based on Equation 9.2.

$$BSI = \frac{\int_0^{\delta_i} F \, dx}{\delta_i t \, J_{Ic}} \tag{9.2}$$

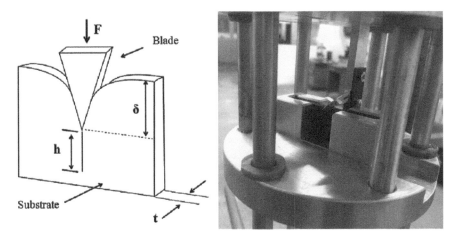

FIGURE 9.3 Schematic illustration of the indentation-type substrate cutting process (left) and experimental set-up within cutting rig (right).

where F, dx, δ_i, t, and J_{Ic} are the applied force, addition of blade displacement, initial depth of blade indentation prior to substrate fracture, thickness of the substrate, and fracture toughness of the substrate, respectively. The typical BSI value ranges from 0.2 for a sharp blade to 0.5 for a blunt blade.

Several studies have provided BSI data on BMG and MGTF-coated blades in comparison with the commercial blades. Tsai et al. conducted a cutting test using $Zr_{48}Cu_{35.3}Al_8Ag_8Si_{0.7}$ BMG blade and 200 nm- $Zr_{48}Cu_{35.3}Al_8Ag_8Si_{0.7}$ -coated MSS commercial blade using no. 11-type surgical blade through styrene butadiene rubber (SBR) substrate. Their result showed that both ZrCuAlAgSi BMG and ZrCuAlAgSi MGTF blades performed low BSI value of 0.25 and 0.23, lower than that on MSS commercial blade (0.34) (Tsai et al., 2012). The similar author also applied blade making on the $Fe_{41}Cr_{15}Mo_{14}C_{12}B_9Y_2Co_7$ BMG and the cutting test revealed the superior sharpness of Fe-based BMG blade (BSI 0.264) than the commercial blade (BSI 0.312). Moreover, the Fe-based blade still remains a lower BSI of 0.376 than the commercial blade (0.596) after 50 mm cutting test (Tsai et al., 2014). Using the similar composition of $Fe_{41}Cr_{15}Mo_{14}C_{12}B_9Y_2Co_7$ MG, Jang et al. fabricated an MGTF-coated layer on the commercial blade and the BSI value indicated the significant sharpness improvement from its bare commercial one (Jang et al., 2015). The BSI of 200 nm-thick Fe-based MGTF-coated blade was at 0.28, which was much lower than that on commercial blade of 0.31 under low indentation rate of 0.08 mm/s and the lower BSI was still resisted even after 30-mm-length cutting with 0.4 of MGTF compared to 0.52 of commercial one.

Tsai et al. once again studied the effect of coating thickness of $(Zr_{48}Cu_{36}Al_8Ag_8)_{9}$ $_{9.5}Si_{0.5}$ (named as Zr48) and $(Zr_{53}Cu_{30}Ni_9Al_8)_{99.5}Si_{0.5}$ (named as Zr53) MGTFs on the cutting sharpness and durability (Tsai et al., 2016). The coating thickness ranged from 200 to 500 nm and the cutting test was conducted at various cutting lengths of 0 to 25 mm. The results indicated that BSI increased along with the addition of cutting length in all blades; however, 200 nm-thick-MGTF of both compositions showed the

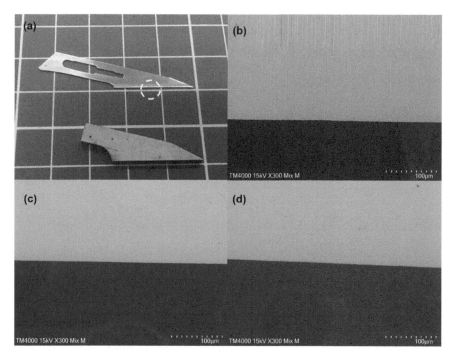

FIGURE 9.4 (a) Commercial CS and ZrCoAl BMG blades with circle indicating the edge-tip region for SEM observation, and initial edge-tip condition for (b) commercial CS, (c) ZrCoAl BMG, and (d) BMGC with 300× of magnification.

lower initial BSI and BSI after 20 cm cutting with 0.255 (Zr48) and 0.247 (Zr53) for initial BSI and 0.441 (Zr48) and 0.437 (Zr53) after 20-cm cutting compared with those on the initial BSI (0.340) and BSI after 20-cm cutting (0.485). Such results also indicated the sharpness improvement on Zr48 and Zr53 MGTF coating of 26% and 23%, respectively, from the commercial blade as reference. Not only for surgical blade, the sharpness improvement was also reported for dermatome made by BMG. Liao et al. fabricated $Fe_{41}C_{10}B_{7.5}Ta_{3.5}$ BMGC to be a skin grafting dermatome and the test on simulated skin showed the sharper edge-tip with a BSI value of 0.35 compared to 0.59 for commercial dermatome (Liao et al., 2020). Additionally, MGTF-coating of $Zr_{53}Cu_{33}Al_9Ta_5$ was also applied on the commercial dermatome substrate and it successfully decreased the BSI value from 0.39 (uncoated) to 0.29, indicating the sharpness improvement of 27% (Chu et al., 2014). The Zr-Co-Al BMG blade was also successfully fabricated in ongoing research using 40 x 20 x 2 BMG plates, resulting in higher quality of cutting edge-tip compared with the crystalline CS commercial blade as shown in Figure 9.4.

9.5.2 Scratch Test for MGTF Adhesion Analysis

The scratch test measures the scratch resistance and substrate adhesion of the coatings. The critical load (Lc), which indicates scratch resistance, is used to evaluate the thin film's adhesion failure at normal load. When the Lc is higher, the applied force to make a scratch on the film is also higher, implying that the coating is more scratch resistant. Lc1 denotes the occurrence of the initial crack, Lc2 (first delamination) is caused by local interfacial spallation or fragmental failure, and Lc3 (total delamination) denotes total coating damage (Thanka Rajan et al., 2016). The work/capability of adhesion was calculated by using Equation 9.3 (Subramanian, 2015):

$$Lc = \frac{\pi d_c^2}{8} \sqrt{\frac{2EW}{t}} \tag{9.3}$$

where Lc, d_c, E, T, and W are the critical load, track width, coating's elastic modulus, film thickness, and work of adhesion, respectively.

Several researchers have reported the scratch test results for respective MGTF coating on the commercial blades. Initially, the industrial requirement of film adhesion force is exceeded 50 N and several MGTFs showed sufficient adhesion to fulfill such an industrial standard. For example, $Fe_{41}Cr_{15}Mo_{14}C_{12}B_9Y_2Co_7$ MGTF performed a strong adhesion after scratch using 100 N of force (Jang et al., 2015). In parallel, $Zr_{48}Cu_{35.3}Al_8Ag_8Si_{0.7}$ MGTF can perform the detached load of 60 N (Tsai et al., 2012). Good adhesion result was also reported by Chu et al. with $Zr_{53}Cu_{33}Al_9Ta_5$ MGTF having detachment at a load of 100 N (Chu et al., 2014). The maximum loading of 65 N was reached with only a negligible delamination of the coating in other Zr-based MGTFs (Chuang et al., 2013). Additionally, by providing several thicknesses in Zr48 and Zr35 MGTFs, Tsai et al. suggested that more coating thickness results in higher adhesion force, with maximum forces being found in 500-nm-thick coating with 87 N (Zr48 MGTF) and 98 N (Zr53 MGTF) (Tsai et al., 2016). The whole BSI and adhesion force data for biocompatible MGs are listed in Table 9.1.

TABLE 9.1
Summary of BSI and Adhesion Force Values on Biocompatible MGs

MGs Composition	Form*	Blade Type	BSI	Adhesion Force (N)	Film Thickness (nm)	Ref.
Commercial SS	-	Surgical	0.31–0.34	>50*	-	(Tsai et al., 2012)
		Dermatome	0.39–0.59	>50*	-	(Chu et al., 2014; Liao et al., 2020)
$Zr_{48}Cu_{35.3}Al_8Ag_8Si_{0.7}$	BMG	Surgical	0.25	-	-	(Tsai et al., 2012)
	MGTF	Surgical	0.23	60	200	(Tsai et al., 2012)
$(Zr_{48}Cu_{36}Al_8Ag_8)_{99.5}Si_{0.5}$	MGTF	Surgical	0.255	50	-	(Tsai et al., 2016)
$(Zr_{53}Cu_{30}Ni_9Al_8)_{99.5}Si_{0.5}$	MGTF	Surgical	0.247	66	200	(Tsai et al., 2016)
$Zr_{53}Cu_{33}Al_9Ta_5$	MGTF	Dermatome	0.29	65	200	(Chu et al., 2014)
$Fe_{41}Cr_{15}Mo_{14}C_{12}B_9Y_2Co_7$	BMG	Surgical	0.264	-	-	(Tsai et al., 2014)
	MGTF	Surgical	0.28	100	200	(Jang et al., 2015)
$Fe_{41}C_{10}B_{7.5}Ta_{3.5}$	BMGC	Dermatome	0.35	-	-	(Liao et al., 2020)

BMG: Bulk Metallic Glass, BMGC: Bulk Metallic Glass Composite, MGTF: Metallic Glass Thin Film.

* The industrial requirement of film adhesion force is >50 N.

FIGURE 9.5 Edge-tip evolution for (a) commercial CS, (b) ZrCoAl BMG, and (c) BMGC after 2× cutting (20-mm-length incision) with 300× magnification indicating blade durability.

9.5.3 BLADE DURABILITY

In clinical surgery, a better sharpness and durability of either surgery blades or dermatomes are able to make a clean and long distance cut with low friction and create a neat wound, resulting in quick recovery. Durability is defined as the ability of a blade to retain the initial condition without any significant functional loss or physical change after a certain period or length of cutting. Based on the informations described above, MGs are believed as a better choice for surgical tool and biomedical part applications due to their durability considering their excellence mechanical properties. In the case of a BMG surgical blade, the durability can be assessed by considering the morphological change of edge-tip after certain length of cutting and a series of wear experiences. This morphological condition of a blade is highly related to the sharpness (BSI value) and surface roughness of the blade, tissue condition after cutting, and the blade biocorrosion resistance. For MGTF, an investigation on the condition of film adhesive after several/certain length of cutting is also needed to indicate the durability. Several studies have been conducted to investigate the durability of BMGs and MGTFs in comparison with the commercial blades (Chu et al., 2014; Jang et al., 2015; Li & Zheng, 2016; Liao et al., 2020; Rajan & Arockiarajan, 2021; Tsai et al., 2012, 2014). Most of them illustrated that for MGs, the surface condition only undergoes a small change and has higher ability to maintain the edge-tip condition after a certain length of cutting, as shown in Figure 9.5. It is different with the commercial blade suffering from gradual deterioration in edge-tip flatness and roughness, which limits them for subsequent application with similar performance to the initial condition. The durability is also represented by the change in BSI values of blades after cutting utilization, with the larger BSI change being noted in commercial surgical blades than in BMG and MGTF-coated blades (Jang et al., 2015; Tsai et al., 2014, 2016).

9.6 CONCLUSION AND FUTURE CHALLENGES

For the last three decades, studies regarding the development and application of MGs have been extensively enriched by the perspective of many researchers on advanced materials due to their unique and superior mechanical, physical, chemical, and thermal properties. Moreover, by excluding the toxic elements such as Be and Ni as well as reducing the content of Cu in the MG alloy systems, MGs are considered as biomaterials having biocompatibility that enables to optimize their excellences as implantable devices and surgical tools. The need of high sharpness, durability, and biocompatibility on surgical blade and dermatome has been answered by the application of several biocompatible MG materials in the form of bulk and thin films coating, and the performance is clearly demonstrated by many researchers indicating the materials' capability as replacement for current conventional crystalline surgical blades. Based on the research, biocompatible BMG and MGTF-coated blades showed a significantly higher sharpness and durability than the commercial ones to provide a reliable incision and neat wound in order to establish faster recovery. However, future challenges come up in developing these advanced surgical technologies. The synthesis and characterization of MGs as well as the fabrication of such metastable, high-strength materials are remained difficult and costly. Considering these issues, the development of low-cost processing technologies is urgently needed to expand the MGs applications, especially for companies or research centers that have no access to such precious equipment, such as vacuum arc melting, electron microscopes, or even nonconventional machining equipment.

REFERENCES

Bai, M.-Y., Chang, Y.-C., & Chu, J. P. (2020). Preclinical studies of non-stick thin film metallic glass-coated syringe needles. *Scientific Reports*, *10*(1), 20313. https://doi.org/10.1038/s41598-020-77008-y

Bakkal, M. (2009). Electron microscopy of bulk metallic glass machining chips. *Journal of Non-Crystalline Solids*, *355*(45–47), 2220–2223. https://doi.org/10.1016/j.jnoncrysol.2009.07.018

Bui, F. M., Mintchev, M. P. & Bott, K. (2000). A quantitative study of the pixel-shifting, blurring and nonlinear distortions in MRI images caused by the presence of metal implants. *Journal of Medical Engineering & Technology*, *24*(1), 20–27. https://doi.org/10.1080/030919000294003

Cai, C.-N., Zhang, C., Sun, Y.-S., Huang, H.-H., Yang, C., & Liu, L. (2017). ZrCuFeAlAg thin film metallic glass for potential dental applications. *Intermetallics*, *86*, 80–87. https://doi.org/10.1016/j.intermet.2017.03.016

Calin, M., Gebert, A., Ghinea, A. C., Gostin, P. F., Abdi, S., Mickel, C., & Eckert, J. (2013). Designing biocompatible Ti-based metallic glasses for implant applications. *Materials Science and Engineering C*, *33*(2), 875–883. https://doi.org/10.1016/j.msec.2012.11.015

Chang, J.-C., Lee, J.-W., Lou, B.-S., Li, C.-L., & Chu, J. P. (2015). Effects of tungsten contents on the microstructure, mechanical and anticorrosion properties of Zr–W–Ti thin film metallic glasses. *Thin Solid Films*, *584*, 253–256. https://doi.org/10.1016/j.tsf.2015.01.063

Charles H. Williamson, M. (2007). Diamond knives: are they the clear choice for clear corneal cataract surgery. *Cataract Refract Surg Today*, *1*, June, 82–84.

Chen, L.-T., Lee, J.-W., Yang, Y.-C., Lou, B.-S., Li, C.-L., & Chu, J. P. (2014). Microstructure, mechanical and anti-corrosion property evaluation of iron-based thin film metallic glasses. *Surface and Coatings Technology*, *260*, 46–55. https://doi.org/10.1016/j.surfcoat.2014.07.039

Chen, X., Xiao, J., Zhu, Y., Tian, R., Shu, X., & Xu, J. (2017). Micro-machinability of bulk metallic glass in ultra-precision cutting. *Materials and Design*, *136*, 1–12. https://doi.org/10.1016/j.matdes.2017.09.049

Chu, J. P., Diyatmika, W., Tseng, Y. J., Liu, Y. K., Liao, W. C., Chang, S. H., Chen, M. J., Lee, J. W., & Jang, J. S. C. (2019). Coating cutting blades with thin-film metallic glass to enhance sharpness. *Scientific Reports*, 9(1), 1–11. https://doi.org/10.1038/s41598-019-52054-3

Chu, J. P., Liu, T. Y., Li, C. L., Wang, C. H., Jang, J. S. C., Chen, M. J., Chang, S. H., & Huang, W. C. (2014). Fabrication and characterizations of thin film metallic glasses: antibacterial property and durability study for medical application. *Thin Solid Films*, *561*, 102–107. https://doi.org/10.1016/j.tsf.2013.08.111

Chu, J. P., Yu, C.-C., Tanatsugu, Y., Yasuzawa, M., & Shen, Y.-L. (2016). Non-stick syringe needles: beneficial effects of thin film metallic glass coating. *Scientific Reports*, 6, 1–7. https://doi.org/10.1038/srep31847

Chuang, C.-Y., Liao, Y.-C., Lee, J.-W., Li, C.-L., Chu, J. P., & Duh, J.-G. (2013). Electrochemical characterization of Zr-based thin film metallic glass in hydrochloric aqueous solution. *Thin Solid Films*, *529*, 338–341.

Etiemble, A., Der Loughian, C., Apreutesei, M., Langlois, C., Cardinal, S., Pelletier, J. M., Pierson, J.-F., & Steyer, P. (2017). Innovative Zr-Cu-Ag thin film metallic glass deposed by magnetron PVD sputtering for antibacterial applications. *Journal of Alloys and Compounds*, *707*, 155–161. https://doi.org/10.1016/j.jallcom.2016.12.259

Fang, H., Hui, X., & Chen, G. (2008). Effects of Mn addition on the magnetic property and corrosion resistance of bulk amorphous steels. *Journal of Alloys and Compounds*, *464*(1), 292–295. https://doi.org/10.1016/j.jallcom.2007.09.139

Gong, P., Deng, L., Jin, J., Wang, S., Wang, X., & Yao, K. (2016). Review on the research and development of Ti-based bulk metallic glasses. *Metals*, *6*(11), 264. https://doi.org/10.3390/met6110264

Gowthaman, S., Balamurugan, K., Kumar, P. M., Ali, S. K. A., Kumar, K. L. M., & Ram Gopal, N. V. (2018). Electrical discharge machining studies on Monel-Super Alloy. *Procedia Manufacturing*, *20*, 386–391. https://doi.org/10.1016/j.promfg.2018.02.056

Gu, X. J., Poon, S. J., & Shiflet, G. J. (2007a). Effects of carbon content on the mechanical properties of amorphous steel alloys. *Scripta Materialia*, *57*(4), 289–292. /https://doi.org/10.1016/j.scriptamat.2007.05.006

Gu, X. J., Poon, S. J., & Shiflet, G. J. (2007b). Mechanical properties of iron-based bulk metallic glasses. *Journal of Materials Research*, *22*(2), 344–351. https://doi.org/DOI: 10.1557/jmr.2007.0036

Han, K., Jiang, H., Wang, Y., Qiang, J., & Yu, C. (2021). Antimicrobial Zr-based bulk metallic glasses for surgical devices applications. *Journal of Non-Crystalline Solids*, *564*(November 2020), 120827. https://doi.org/10.1016/j.jnoncrysol.2021.120827

Hua, N., Huang, L., Chen, W., He, W., & Zhang, T. (2014). Biocompatible Ni-free Zr-based bulk metallic glasses with high-Zr-content: compositional optimization for potential biomedical applications. *Materials Science and Engineering C*, *44*, 400–410. https://doi.org/10.1016/j.msec.2014.08.049

Hua, N., Huang, L., Wang, J., Cao, Y., He, W., Pang, S., & Zhang, T. (2012). Corrosion behavior and *in vitro* biocompatibility of Zr-Al-Co-Ag bulk metallic glasses: an experimental case study. *Journal of Non-Crystalline Solids*, *358*(12–13), 1599–1604. https://doi.org/10.1016/j.jnoncrysol.2012.04.022

Huang, C. H., Huang, J. C., Li, J. B., & Jang, J. S. C. (2013). Simulated body fluid electrochemical response of Zr-based metallic glasses with different degrees of crystallization. *Materials Science and Engineering C*, *33*(7), 4183–4187. https://doi.org/10.1016/j.msec.2013.06.007

Huang, H., & Yan, J. (2015). On the surface characteristics of a Zr-based bulk metallic glass processed by microelectrical discharge machining. *Applied Surface Science*, *355*, 1306–1315. https://doi.org/10.1016/j.apsusc.2015.08.239

Huang, H., & Yan, J. (2016). Microstructural changes of Zr-based metallic glass during micro-electrical discharge machining and grinding by a sintered diamond tool. *Journal of Alloys and Compounds*, *688*, 14–21. https://doi.org/10.1016/j.jallcom.2016.07.181

Huang, J. C., Chu, J. P., & Jang, J. S. C. (2009). Recent progress in metallic glasses in Taiwan. *Intermetallics*, *17*(12), 973–987. https://doi.org/10.1016/j.intermet.2009.05.004

Inoue, A. (2000). Stabilization of metallic supercooled liquid. *Acta Mater.*, *48*, 279–306.

Inoue, A., & Takeuchi, A. (2011). Recent development and application products of bulk glassy alloys. *Acta Materialia*, *59*(6), 2243–2267. https://doi.org/10.1016/j.actamat.2010.11.027

Inoue, A., & Zhang, T. (1996). Fabrication of bulk glassy $Zr_{55}Al_{10}Ni_5Cu_{30}$ alloy of 30 mm in diameter by a suction casting method. *Material Transitions*, *37*, 185–187.

Inoue, A., Zhang, T., Nishiyama, N., Ohba, K., & Masumoto, T. (1993). Preparation of 16 mm diameter rod of amorphous Zr65Al7.5Ni10Cu17.5 alloy. *Materials Transactions, JIM*, *34*(12), 1234–1237.

Jabed, A., Khan, M. M., Camiller, J., Greenlee-Wacker, M., Haider, W., & Shabib, I. (2019). Property optimization of Zr-Ti-X (X = Ag, Al) metallic glass via combinatorial development aimed at prospective biomedical application. *Surface and Coatings Technology*, *372*, 278–287. https://doi.org/10.1016/j.surfcoat.2019.05.036

Jang, J. S., Tsai, P., Shiao, A., Li, T., Chen, C., Peter, J., Duh, J., Chen, M., & Chang, S. (2015). Enhanced cutting durability of surgical blade by coating with Fe-based metallic glass thin fi lm. *Intermetallics*, *65*, 56–60. https://doi.org/10.1016/j.intermet.2015.06.012

Ke, J. L., Huang, C. H., Chen, Y. H., Tsai, W. Y., Wei, T. Y., & Huang, J. C. (2014). *In vitro* biocompatibility response of Ti–Zr–Si thin film metallic glasses. *Applied Surface Science*, *322*, 41–46. https://doi.org/10.1016/j.apsusc.2014.09.204

Klement Jun., W., Willens, R. H., & Duwez, P. O. L. (1960). Non-crystalline structure in solidified gold–silicon alloys. *Nature*, *187*, 869. https://doi.org/10.1038/187869b0

Koyano, T., Sugata, Y., Hosokawa, A., & Furumoto, T. (2019). Micro-electrical discharge machining of micro-rods using tool electrodes with high electrical resistivity. *Precision Engineering*, *55*, 95–100. https://doi.org/10.1016/j.precisioneng.2018.08.013

Krejcie, A. J., Kapoor, S. G., & Devor, R. E. (2012). A hybrid process for manufacturing surgical-grade knife blade cutting edges from bulk metallic glass. *Journal of Manufacturing Processes*, *14*(1), 26–34. https://doi.org/10.1016/j.jmapro.2011.09.001

Kumar, J. (2013). Ultrasonic machining—a comprehensive review. *Machining Science and Technology*, *17*(3), 325–379. https://doi.org/10.1080/10910344.2013.806093

Kuriakose, S., Patowari, P. K., & Bhatt, J. (2017). Machinability study of Zr-Cu-Ti metallic glass by micro hole drilling using micro-USM. *Journal of Materials Processing Technology*, *240*, 42–51. https://doi.org/10.1016/j.jmatprotec.2016.08.026

Li, H., Zheng, Y., & Qin, L. (2014). Progress of biodegradable metals. *Progress in Natural Science*, *24*(5), 414–422. https://doi.org/10.1016/j.pnsc.2014.08.014

Li, H. F., & Zheng, Y. F. (2016). Recent advances in bulk metallic glasses for biomedical applications. *Acta Biomaterialia*, *36*, 1–20. https://doi.org/10.1016/j.actbio.2016.03.047

Li, H. F., Zheng, Y. F., Xu, F., & Jiang, J. Z. (2012). *In vitro* investigation of novel Ni free Zr-based bulk metallic glasses as potential biomaterials. *Materials Letters*, *75*, 74–76. https://doi.org/10.1016/j.matlet.2012.01.105Li, J., Shi, L., Zhu, Z., He, Q., Ai, H., & Xu, J. (2013). $Zr_{61}Ti_2Cu_{25}Al_{12}$ metallic glass for potential use in dental implants: biocompatibility assessment by *in vitro* cellular responses. *Materials Science and Engineering: C*, *33*(4), 2113–2121. https://doi.org/10.1016/j.msec.2013.01.033

Li, T. H., Liao, Y. C., Song, S. M., Jiang, Y. L., Tsai, P. H., Jang, J. S. C., & Huang, J. C. (2018). Significantly enhanced mechanical properties of ZrAlCo bulk amorphous alloy by microalloying with Ta. *Intermetallics*, *93*(November 2017), 162–168. https://doi.org/10.1016/j.intermet.2017.12.008

Li, T. H., Tsai, P. H., Hsu, K. T., Liu, Y. C., Jang, J. S. C., & Huang, J. C. (2016). Significantly enhanced drilling ability of the orthopedic drill made of Zr-based bulk metallic glass composite. *Intermetallics*, *78*, 17–20. https://doi.org/10.1016/j.intermet.2016.08.001

Li, T. H., Wong, P. C., Chang, S. F., Tsai, P. H., Jang, J. S. C., & Huang, J. C. (2017). Biocompatibility study on Ni-free Ti-based and Zr-based bulk metallic glasses. *Materials Science and Engineering: C*, *75*, 1–6. https://doi.org/10.1016/j.msec.2017.02.006

Li, Z., Zhang, C., & Liu, L. (2015). Wear behavior and corrosion properties of Fe-based thin film metallic glasses. *Journal of Alloys and Compounds*, *650*, 127–135. https://doi.org/10.1016/j.jallcom.2015.07.256

Liao, Y. C., Song, S. M., Li, T. H., Tsai, P. H., Chen, C. Y., Jang, J. S. C., Chu, J. P., & Tseng, C. M. (2020). Enhanced toughness and skin grafting durability of a dermatome made of Fe-based bulk metallic glass composites. *Materials Chemistry and Physics, 241,* 12228. https://doi.org/10.1016/j.matchemphys.2019.122281

Lin, C. H., Huang, C. H., Chuang, J. F., Huang, J. C., Jang, J. S. C., & Chen, C. H. (2013). Rapid screening of potential metallic glasses for biomedical applications. *Materials Science & Engineering. C, Materials for Biological Applications*, *33*(8), 4520–4526. https://doi.org/10.1016/j.msec.2013.07.006

Lin, H. C., Tsai, P. H., Ke, J. H., Li, J. B., Jang, J. S. C., Huang, C. H., & Haung, J. C. (2014). Designing a toxic-element-free Ti-based amorphous alloy with remarkable supercooled liquid region for biomedical application. *Intermetallics*, *55*, 22–27. https://doi.org/10.1016/j.intermet.2014.07.003

Liu, L., Qiu, C. L., Chen, Q., & Zhang, S. M. (2006). Corrosion behavior of Zr-based bulk metallic glasses in different artificial body fluids. *Journal of Alloys and Compounds*, *425*(1), 268–273. https://doi.org/10.1016/j.jallcom.2006.01.048

Liu, L., Qiu, C. L., Huang, C. Y., Yu, Y., Huang, H., & Zhang, S. M. (2009). Biocompatibility of Ni-free Zr-based bulk metallic glasses. *Intermetallics*, *17*(4), 235–240. https://doi.org/10.1016/j.intermet.2008.07.022

Liu, Z., Chan, K. C., & Liu, L. (2011). Development of Ni- and Cu-Free Zr-based bulk metallic glasses for biomedical applications. *Materials Transactions*, *52*(1), 61–67. https://doi.org/10.2320/matertrans.M2010068

Lou, B.-S., Lin, T.-Y., Chen, W.-T., & Lee, J.-W. (2019). Corrosion property and biocompatibility evaluation of Fe–Zr–Nb thin film metallic glasses. *Thin Solid Films*, *691*, 137615. https://doi.org/10.1016/j.tsf.2019.137615

Maroju, N. K., & Jin, X. (2018). Vibration-assisted dimple generation on bulk metallic glass. *Procedia Manufacturing*, *26*, 317–328. https://doi.org/10.1016/j.promfg.2018.07.040

Maroju, N. K., Yan, D. P., Xie, B., & Jin, X. (2018). Investigations on surface microstructure in high-speed milling of Zr-based bulk metallic glass. *Journal of Manufacturing Processes*, *35*(June), 40–50. https://doi.org/10.1016/j.jmapro.2018.07.020

McCarthy, C. T., Annaidh, A. N., & Gilchrist, M. D. (2010). On the sharpness of straight edge blades in cutting soft solids: Part II—Analysis of blade geometry. *Engineering Fracture Mechanics*, *77*(3), 437–451. https://doi.org/10.1016/j.engfracmech.2009.10.003

McCarthy, C. T., Hussey, M., & Gilchrist, M. D. (2007). On the sharpness of straight edge blades in cutting soft solids: Part I—indentation experiments. *Engineering Fracture Mechanics*, *74*(14), 2205–2224. https://doi.org/10.1016/j.engfracmech.2006.10.015

Meagher, P., O'Cearbhaill, E. D., Byrne, J. H., & Browne, D. J. (2016). Bulk metallic glasses for implantable medical devices and surgical tools. *Advanced Materials*, *28*(27), 5755–5762. https://doi.org/10.1002/adma.201505347

Meng, L., Zeng, Y., & Zhu, D. (2017). Micropatterning of Ni-based metallic glass by pulsed wire electrochemical micro machining. *Electrochimica Acta*, *233*, 274–283. https://doi.org/10.1016/j.electacta.2017.03.045

Morrison, M. L., Buchanan, R. A., Peker, A., Liaw, P. K., & Horton, J. A. (2007). Electrochemical behavior of a Ti-based bulk metallic glass. *Journal of Non-Crystalline Solids*, *353*(22), 2115–2124. https://doi.org/10.1016/j.jnoncrysol.2007.03.012

Morrison, M. L., Buchanan, R. A., Peker, A., Peter, W. H., Horton, J. A., & Liaw, P. K. (2004). Cyclic-anodic-polarization studies of a $Zr_{41.2}Ti_{13.8}Ni_{10}Cu_{12.5}Be_{22.5}$ bulk metallic glass. *Intermetallics*, *12*(10), 1177–1181. https://doi.org/10.1016/j.intermet.2004.04.005

Nguyen, V. T., Li, T. H., Song, S. M., Liao, Y. C., Tsai, P. H., Wong, P. C., Nguyen, V. C., & Jang, J. S. C. (2019). Synthesis of biocompatible TiZr-based bulk metallic glass foams for bio-implant application. *Materials Letters*, *256,* 126650. https://doi.org/10.1016/j.matlet.2019.126650

Nguyen, V. T., Wong, X. P. C., Song, S. M., Tsai, P. H., Jang, J. S. C., Tsao, I. Y., Lin, C. H., & Nguyen, V. C. (2020). Open-cell tizr-based bulk metallic glass scaffolds with excellent biocompatibility and suitable mechanical properties for biomedical application. *Journal of Functional Biomaterials*, *11*(2), 28. https://doi.org/10.3390/jfb11020028

Ni, H. S., Liu, X. H., Chang, X. C., Hou, W. L., Liu, W., & Wang, J. Q. (2009). High performance amorphous steel coating prepared by HVOF thermal spraying. *Journal of Alloys and Compounds*, *467*(1), 163–167. https://doi.org/10.1016/j.jallcom.2007.11.133

Nishiyama, N., Takenaka, K., Miura, H., Saidoh, N., Zeng, Y., & Inoue, A. (2012). The world's biggest glassy alloy ever made. *Intermetallics*, *30*, 19–24. https://doi.org/10.1016/j.intermet.2012.03.020

Oak, J.-J., Hwang, G.-W., Park, Y.-H., Kimura, H., Yoon, S.-Y., & InouE, A. (2009). Characterization of Surface properties, osteoblast cell culture *in vitro* and processing with flow-viscosity of Ni-free Ti-based bulk metallic glass for biomaterials. *Journal of Biomechanical Science and Engineering*, *4*(3), 384–391. https://doi.org/10.1299/jbse.4.384

Oak, J.-J., Louzguine-Luzgin, D. V., & Inoue, A. (2007). Fabrication of Ni-free Ti-based bulk-metallic glassy alloy having potential for application as biomaterial, and investigation of its mechanical properties, corrosion, and crystallization behavior. *Journal of Materials Research*, *22*(5), 1346–1353. https://doi.org/10.1557/jmr.2007.0154

Obeydavi, A., Shafyei, A., Rezaeian, A., Kameli, P., & Lee, J.-W. (2020). Microstructure, mechanical properties and corrosion performance of $Fe_{44}Cr_{15}Mo_{14}Co7C_{10}B_5Si_5$ thin film metallic glass deposited by DC magnetron sputtering. *Journal of Non-Crystalline Solids*, *527*, 119718. https://doi.org/10.1016/j.jnoncrysol.2019.119718

Padhy, N., Ningshen, S., & Kamachi Mudali, U. (2010). Electrochemical and surface investigation of zirconium based metallic glass $Zr_{59}Ti_3Cu_{20}Al_{10}Ni_8$ alloy in nitric acid and sodium chloride media. *Journal of Alloys and Compounds*, *503*(1), 50–56. https://doi.org/10.1016/j.jallcom.2010.05.002

Pang, S., Liu, Y., Li, H., Sun, L., Li, Y., & Zhang, T. (2015). New Ti-based Ti – Cu – Zr – Fe – Sn – Si – Ag bulk metallic glass for biomedical applications. *Journal of Alloys and Compounds*, *625*, 323–327. https://doi.org/10.1016/j.jallcom.2014.07.021

Patel, J. D., & Maniya, K. D. (2018). A review on: wire cut electrical discharge machining process for metal matrix composite. *Procedia Manufacturing*, *20*, 253–258.

Peker, A., & Johnson, W. L. (1993). A highly processable metallic glass: $Zr_{41.2}Ti_{13.8}Cu_{12.5}Ni_{10.0}Be_{22.5}$. *Applied Physics Letters*, *63*(17), 2342–2344.

Phan, T. A., Lee, S. M., Makino, A., Oguchi, H., & Kuwano, H. (2011). Fe-B-Nb-Nd magnetic metallic glass thin film for MEMS/NEMS structure. *2011 IEEE 24th International*

Conference on Micro Electro Mechanical Systems, 428–431. https://doi.org/10.1109/MEMSYS.2011.5734453

Ponnambalam, V., Poon, S. J., Shiflet, G. J., Keppens, V. M., Taylor, R., & Petculescu, G. (2003). Synthesis of iron-based bulk metallic glasses as nonferromagnetic amorphous steel alloys. *Applied Physics Letters*, *83*(6), 1131–1133. https://doi.org/10.1063/1.1599636

Pradana, Y. R. A., Ferara, A., Aminnudin, A., Wahono, W., Jang, J. S. C., & Jang, J. S. C. (2020). The effect of discharge current and pulse-on time on biocompatible Zr-based BMG sinking-EDM. *Open Engineering*, *10*(1), 401–407. https://doi.org/10.1515/eng-2020-0049

Rajan, S. T., & Arockiarajan, A. (2021). Thin film metallic glasses for bioimplants and surgical tools: a review. *Journal of Alloys and Compounds*, *876*, 159939. https://doi.org/10.1016/j.jallcom.2021.159939

Shafiei, F., Honda, E., Takahashi, H., & Sasaki, T. (2003). Artifacts from dental casting alloys in magnetic resonance imaging. *Journal of Dental Research*, *82*(8), 602–606. https://doi.org/10.1177/154405910308200806

Subramanian, B. (2015). *In vitro* corrosion and biocompatibility screening of sputtered $Ti_{40}Cu_{36}Pd_{14}Zr_{10}$ thin film metallic glasses on steels. *Materials Science and Engineering: C*, *47*, 48–56. https://doi.org/10.1016/j.msec.2014.11.013

Subramanian, B., Maruthamuthu, S., & Rajan, S. T. (2015). Biocompatibility evaluation of sputtered zirconium-based thin film metallic glass-coated steels. *International Journal of Nanomedicine*, *10*(Suppl 1), 17–29. https://doi.org/10.2147/IJN.S79977

Sun, Y., Huang, Y., Fan, H., Liu, F., Shen, J., Sun, J., & Chen, J. J. J. (2014). Comparison of mechanical behaviors of several bulk metallic glasses for biomedical application. *Journal of Non-Crystalline Solids*, *406*, 144–150. https://doi.org/10.1016/j.jnoncrysol.2014.09.021

Thanka Rajan, S., Bendavid, A., & Subramanian, B. (2019). Cytocompatibility assessment of Ti-Nb-Zr-Si thin film metallic glasses with enhanced osteoblast differentiation for biomedical applications. *Colloids and Surfaces B: Biointerfaces*, *173*, 109–120. https://doi.org/10.1016/j.colsurfb.2018.09.041

Thanka Rajan, S., Karthika, M., Bendavid, A., & Subramanian, B. (2016). Apatite layer growth on glassy $Zr_{48}Cu_{36}Al_8Ag_8$ sputtered titanium for potential biomedical applications. *Applied Surface Science*, *369*, 501–509. https://doi.org/10.1016/j.apsusc.2016.02.054

Thanka Rajan, S., Nandakumar, A. K., Hanawa, T., & Subramanian, B. (2017). Materials properties of ion beam sputtered Ti-Cu-Pd-Zr thin film metallic glasses. *Journal of Non-Crystalline Solids*, *461*, 104–112. https://doi.org/10.1016/j.jnoncrysol.2017.01.008

Tsai, P. H., Li, T. H., Hsu, K. T., Chiou, J. W., Jang, J. S. C., & Chu, J. P. (2016). Effect of coating thickness on the cutting sharpness and durability of Zr-based metallic glass thin film coated surgical blades. *Thin Solid Films*, *618*, 36–41. https://doi.org/10.1016/j.tsf.2016.05.020

Tsai, P. H., Lin, Y. Z., Li, J. B., Jian, S. R., Jang, J. S. C., Li, C., Chu, J. P., & Huang, J. C. (2012). Sharpness improvement of surgical blade by means of ZrCuAlAgSi metallic glass and metallic glass thin film coating. *Intermetallics*, *31*, 127–131. https://doi.org/10.1016/j.intermet.2012.06.014

Tsai, P. H., Xiao, A. C., Li, J. B., Jang, J. S. C., Chu, J. P., & Huang, J. C. (2014). Prominent Fe-based bulk amorphous steel alloy with large supercooled liquid region and superior corrosion resistance. *Journal of Alloys and Compounds*, *586*, 94–98. https://doi.org/10.1016/j.jallcom.2013.09.186

Wada, T., Qin, F., Wang, X., Yoshimura, M., Inoue, A., Sugiyama, N., Ito, R., & Matsushita, N. (2009). Formation and bioactivation of Zr-Al-Co bulk metallic glasses. *Journal of Materials Research*, *24*(9), 2941–2948. https://doi.org/10.1557/jmr.2009.0348

Wang, K., Zhang, Q., & Zhang, J. (2019). Evaluation of scale effect of micro electrical discharge machining system. *Journal of Manufacturing Processes*, *38*, 174–178. https://doi.org/10.1016/j.jmapro.2019.01.005

Wang, Y. B., Li, H. F., Cheng, Y., Wei, S. C., & Zheng, Y. F. (2009). Corrosion performances of a nickel-free Fe-based bulk metallic glass in simulated body fluids. *Electrochemistry Communications*, *11*(11), 2187–2190. https://doi.org/10.1016/j.elecom.2009.09.027

Wang, Y. B., Li, H. F., Zheng, Y. F., & Li, M. (2012). Corrosion performances in simulated body fluids and cytotoxicity evaluation of Fe-based bulk metallic glasses. *Materials Science and Engineering: C*, *32*(3), 599–606. https://doi.org/10.1016/j.msec.2011.12.018

Wang, Y. B., Zheng, Y. F., Wei, S. C., & Li, M. (2011). *In vitro* study on Zr-based bulk metallic glasses as potential biomaterials. *Journal of Biomedical Materials Research Part B: Applied Biomaterials*, *96B*(1), 34–46. https://doi.org/10.1002/jbm.b.31725

Whitesides, G. M., & Wong, A. P. (2006). The intersection of biology and materials science. *MRS Bulletin*, *31*(1), 19–27. https://doi.org/10.1557/mrs2006.2

Williams, E., & Lavery, N. (2017). Laser processing of bulk metallic glass: a review. *Journal of Materials Processing Technology*, *247*(March), 73–91. https://doi.org/10.1016/j.jmatprotec.2017.03.034

Wong, P., Song, S., Tsai, P., Maqnun, M. J., Wang, W., Wu, J., & Jang, S. J. (2022). Using Cu as a spacer to fabricate and control the porosity of titanium zirconium based bulk metallic glass foams for orthopedic implant applications. *Materials, 15*(5), 1887. https://doi.org/10.3390/ma15051887

Yiu, P., Diyatmika, W., Bönninghoff, N., Lu, Y.-C., Lai, B.-Z., & Chu, J. P. (2020). Thin film metallic glasses: properties, applications and future. *Journal of Applied Physics*, *127*(3), 30901. https://doi.org/10.1063/1.5122884

Zhu, G., Zhang, Q., Wang, K., Huang, Y., & Zhang, J. (2018). Effects of different electrode materials on high-speed electrical discharge machining of W9Mo3Cr4V. *Procedia CIRP*, *68*, 64–69. https://doi.org/10.1016/j.procir.2017.12.023

Zhu, S. L., Wang, X. M., & Inoue, A. (2008). Glass-forming ability and mechanical properties of Ti-based bulk glassy alloys with large diameters of up to 1 cm. *Intermetallics*, *16*(8), 1031–1035. https://doi.org/10.1016/j.intermet.2008.05.006

Zohdi, H., Shahverdi, H. R., & Hadavi, S. M. M. (2011). Effect of Nb addition on corrosion behavior of Fe-based metallic glasses in Ringer's solution for biomedical applications. *Electrochemistry Communications*, *13*(8), 840–843. https://doi.org/10.1016/j.elecom.2011.05.017

10 Synthesis and Characterization of Zinc Ferrite as Nanofluid Heat Exchanger Deploying Co-precipitation Method

Poppy Puspitasari,[1,2] Yuke Nofantyu,[1]*
Avita Ayu Permanasari,[1,2] Riana Nurmalasari,[1]
and Andika Bagus Nur Rahma Putra[1]
[1] Department of Mechanical Engineering, Faculty of
Engineering, Universitas Negeri Malang, Indonesia
[2] Centre of Advanced Materials for Renewable Energy
(CAMRY), State University of Malang, Indonesia
[*] Corresponding author: Poppy Puspitasari

CONTENTS

10.1 INTRODUCTION

Nanomaterials are currently one of the most active research areas in physics, chemistry, and engineering. Evidence of this interest is demonstrated by the large number of recent conferences and research papers dedicated to the subject. There are several

DOI: 10.1201/9781003320746-10

reasons for this, one of which is the need to manufacture new materials at a finer scale to continuously reduce costs and increase the speed of information transmission and storage. Another reason is that nanomaterials display novel improved properties compared to conventional materials, which open up possibilities for new technological applications. Nanomaterials are substances that have a grain size of one per millionth of a meter. Nanomaterials manifest attractive and useful extraordinary properties, which can be exploited for the branching of structural and nonstructural packages (Gajanan & Tijare, 2018). Some of the problems that can be raised in nanomaterial research are the ability to control size, the ability to control the possibility of defects that occur, the concentration of the solution used, the atomic interactions that occur, and the synthesis process (Puspitasari, 2017).

Nanocrystalline ferrite is a very attractive material because its unique dielectric, magnetic, and optical properties make it attractive from both a scientific and a technological point of view (Dar, Shah, Siddiqui, & Kotnala, 2014). Spinel ferrite nanoparticles with high surface area have many technical applications in several fields, such as high-density information storage, ferro-fluids, catalysts, drug targets, hyperthermia, magnetic separation, and magnetic resonance imaging (MRI) (Lehyani et al., 2017).

Spinel ferrite has the structural formula MFe_2O_4 (M is a metal ion of 3d transition elements such as Mn, Ni, Cu, Co, Zn, Mg, and Fe) with a spinel cubic crystal structure. Among many spinel ferrites, the study of zinc ferrite ($ZnFe_2O_4$) nanoparticles is one of the most interesting researches. This is due to its unique chemical and thermal stability, as well as the dependence of its magnetic properties on particle size (Asmin, Mutmainnah, & Suharyadi, 2015).

So far, there are several methods to produce ferrite, such as conventional ceramic technique, chemical co-precipitation method, hydrothermal synthesis method, spray pyrolysis method, microemulsion method, and sol-gel process, among which conventional ceramic technique is the main way to produce ferrite in application. Efforts to control the composition precisely and avoid impurities have been made, and wet chemical methods are now attracting the attention of researchers, one of which is the co-precipitation method (Chen, Xia, & Dai, 2015).

The co-precipitation technique is the most suitable technique for the synthesis of $ZnFe_2O_4$ nanoparticles because of the very small and uniform crystal size distribution, simplicity, and the possibility of unnecessary calcination. Based on the theory in co-precipitation technique, several parameters such as counter ion, ionic strength, pH, and precipitation temperature can affect the structure and magnetic properties of ferrite (Milanović et al., 2013). Science and technology in the last few decades have developed rapidly. The development of science and technology is indicated by the presence of quality industrial products employing the latest technology. One of the recent technologies is the use of nanofluids in the cooling system. Nanofluid is a cooling fluid with added nanomaterials and is one of the products of nanotechnology which is currently being developed rapidly. Research on nanofluids has shown that nanofluids have a higher heat transfer value compared to conventional cooling fluids (Firlianda, Permanasari, Puspitasari, & Sukarni, 2019). Nanofluids also have better thermal conductivity than fluids without nanoparticles (Permanasari, Puspitasari, & Sukarni, 2019; Tripathi & Chandra, 2015).

The zinc ferrite nanoparticle material is one of the materials that has been widely studied in recent years. Research on $ZnFe_2O_4$ nanoparticles was carried out because these nanoparticles have unique magnetic, electrical, and photocatalytic properties (Qin et al., 2017). Based on its properties, $ZnFe_2O_4$ nanoparticles can be widely applied as lithium ion batteries, catalysts (Kerroum et al., 2019), solar cells, sensors (Fabbiyola et al., 2015), and nanofluids for radiator coolants (Tripathi & Chandra, 2015).

The zinc ferrite nanoparticle materials can be synthesized using the co-precipitation method (Kerroum et al., 2019), which is widely used for wet chemical processes. The nanomaterial produced by this method has several advantages over other wet chemical methods, namely, high homogeneity, high purity, and easy preparation process. Solution concentration, reaction temperature, and pH value are the main factors affecting the final particle formation result in this method (Qin et al., 2017).

Howon Lee et al. (Lee et al., 2008) conducted research with various pH values ranging from pH 3 to pH12. This study indicates that $ZnFe_2O_4$ is formed in the pH range of 6–12. Previous research conducted by Ait Keroum showed that the synthesis of $ZnFe_2O_4$ with a variation of pH values 9–12 resulted in an increase in crystal size from 19 to 33 nm along with increasing pH values. Unlike the two studies mentioned earlier, this study aimed to analyze the effect of pH value on the synthesis co-precipitation method on the test results of X-ray diffraction (XRD), scanning electron microscope (SEM), and Fourier transform infra-red (FTIR) of $ZnFe_2O_4$ and to determine the effectiveness of $ZnFe_2O_4$ nanofluids compared with conventional fluids.

10.1.1 PREVIOUS RESEARCH

Zinc ferrite oxide ($ZnFe_2O_4$) is a nanoparticle material that has been widely studied by researchers because it has basic properties that can be applied into various fields such as gas sensing devices, lithium-ion batteries, catalysts, and nanomedicine (Kerroum et al., 2019). Recent studies have shown that $ZnFe_2O_4$ magnetic particles can be used as potential agents in MRI or as heat sources in magnetic hyperthermia applications. The bulk $ZnFe_2O_4$ material usually adopts a normal spinel structure where all the non-magnetic Zn2+ cations are located in the tetrahedral site and exhibit paramagnetism at room temperature. Dimensional reduction of $ZnFe_2O_4$ to the nanoparticle scale resulted in Zn2+ cations being distributed together in both tetrahedral and octahedral sites (Vinosha, Mely, Jeronsia, Krishnan, & Das, 2017).

Zinc ferrite oxide ($ZnFe_2O_4$) nanoparticles can be synthesized using various synthesis methods such as sol-gel, ball milling, thermal decomposition, hydrothermal, and co-precipitation. The co-precipitation method was chosen because the process is easy, produces nanoparticles with high purity, and is of low cost. The process of the co-precipitation method is to make a mixed solution of the basic ingredients as precursors and add an alkaline solution to control the pH value. Parameters that can affect the yield of nanoparticles in this process are stirring speed, precursors ratio, reaction temperature, and the pH value of the mixed solution.

The method used to obtain zinc ferrite oxide nanoparticles can use the co-precipitation synthesis method in which zinc ferrite oxide nanoparticles are prepared with materials such as iron (III) chloride hexahydrate ($FeCl_3.6H_2O$), zinc chloride

(ZnCl$_2$), and sodium hydroxide (NaOH) using the following chemical reaction formula (Kerroum et al., 2019):

$$ZnCl_2 + 2FeCl_3 + 8NaOH \rightarrow ZnFe_2O_4 + 4H_2O + 8NaCl$$

MFe$_2$O$_4$ nanoparticles are very important magnetic materials because they have good potential in various modern technological applications where M is the divalent metal ion of the 3d transition element in the periodic table of elements. Various sources have reported that ZnFe$_2$O nanoparticles can be applied to gas sensors, MRI (Kerroum et al., 2019), and photocatalysts (Vinosha et al., 2017).

The crystal structure and the size of ZnFe$_2$O$_4$ nanoparticles synthesized using the co-precipitation method were determined using XRD analysis. The diffraction characteristic peaks for all samples occurred at 2θ equal to 29.89°, 35.21°, 36.84°, 42.79°, 53.08°, 56.58°, and 62.13° with Miller indices (220), (311), (222), (400), (422), (511), and (440), respectively. The diffraction pattern shows that for the four samples the spinel cubic phase corresponds to the Joint Committee on Powder Diffraction Standards (JCPDS) data base no. 22–1012. The XRD pattern of the peak (311) at the 2θ position equal to 35.21o can be used to calculate the average crystal size. Based on the results of calculations using the Scherrer equation, the crystal size increased from 15 to 24 nm as the pH value increased. The increase in crystal size may be attributed to the acceleration of the growth phase which is a consequence of the addition of a larger amount of base substance (Kerroum et al., 2019). Other studies have shown smaller crystal sizes with the use of alkaline substances that have higher concentrations.

The magnetic properties of ZnFe$_2$O$_4$ were shown by magnetization curve with samples Z8009 and Z8012 synthesized with pH values 9 and 12. The magnetization curve at low temperature (5 K) showed that both samples had ferromagnetic behavior with an open hysteresis loop. The magnetization of the two samples was unsaturated when increasing the external static magnetic field to 60 kOe. The magnetization corresponds to the largest external magnetic field (60 kOe) which is 41 emu/g for sample Z8012 and 35 emu/g for sample Z8009. The coercivity values at 5 K for sample Z8009 are 739 Oe and 300 Oe for sample Z012. Unexpectedly, the largest anisotropy (Hc coercive field) comes from the smallest particles. This could be explained by the high magnetic dipole interactions and could reveal the presence of swarms of nanoparticles (Kerroum et al., 2019). The coercive field Hc and the remanent magnetization Mr at 300 K are equal to zero for both samples, indicating that the sample behaves superparamagnetically. The maximum magnetization of both samples dropped significantly to a value of 12 emu/g, probably due to the spin canting effect on the surface which can be thought of as a magnetic shell surrounding the magnetic core (Kerroum et al., 2019).

10.1.2   HEAT TRANSFER

Heat transfer is a branch of science that studies and predicts how heat energy can move from one place to another. The absolute requirement for heat transfer is that there is a temperature difference between one part and another (Holman, 2010). Heat will move from a place that has a high temperature to a place that has a lower temperature; in other words, an object with a high temperature will release heat and an object with a low temperature will receive heat. The principle of heat transfer can be applied to a device commonly called a heat exchanger. A heat exchanger is a device that utilizes temperature differences for heat transfer between two objects without any mass transfer. The tool can function as a heater or cooler depending on the needs needed.

A simple heat exchanger is usually a double pipe consisting of a tube and a jacket on the outside. Heat transfer that occurs in heat exchangers usually involves convection of the fluid and conduction along the wall that separates the two fluids (Incropera, Dewitt, Bergman, & Lavine, 2007). The most common heat exchanger we encounter is the radiator in motor vehicles. A working motor engine will generate heat from the fuel combustion process and friction between the piston and the engine wall, so a heat exchanger in the form of a radiator is needed to cool the engine so that it does not overheat. The radiator utilizes fluid as a coolant where the fluid will be passed to the hot engine wall to cool the engine continuously. Based on the above description, the heat transfer system in the heat exchanger has a temperature difference between the hot object and the cold fluid; using this temperature difference one can calculate the value of heat transfer efficiency and the value of the logarithmic average temperature difference of the heat exchanger using Equations 10.1 and 10.2 as follows:

$$\eta = \frac{T_1 - T_2}{T_1 - t_1} \times 100\% \qquad (10.1)$$

$$\Delta T_m = \frac{(T_1 - t_1) - (T_2 - t_2)}{\ln \dfrac{T_1 - t_1}{T_2 - t_2}} \qquad (10.2)$$

where
η = Efficiency [%],
ΔT_m = Logarithmic mean of temperature differences [°C],
T_1 = Inlet temperature of hot fluid [°C],
T_2 = Outlet temperature of hot fluid [°C],
t_1 = Inlet temperature of cold fluid [°C], and
t_2 = Outlet temperature of cold fluid [°C].

The calorific value exchanged in the heat transfer system of the heat exchanger can also be determined using Equation 10.3 as follows:

$$q = A \times U \times \Delta T_m \qquad (10.3)$$

where
q = Heat exchanged [kkal/jam],
A = Surface area of the heat exchanger [m²], and
U = Heat transmission capacity [kkal/m²h°C].

10.1.3 CO-PRECIPITATION METHODS

The size of the nanomaterial with the desired dimensions can be achieved by improving the synthesis method. Uniform particles are usually prepared by homogeneous precipitation reactions in the process of nucleation separation and growth of nuclei (Kandpal, Sah, Loshali, Joshi, & Prasad, 2014).

A number of synthesis methods have been developed to meet the demand for magnetic nanoparticles (MNPs) with controlled properties including sol-gel synthesis, hydrothermal reactions, sonochemical procedures, hydrolysis and thermolysis of precursors, electrospray synthesis, and synthesis of microemulsions. However, one common and economical method for the controlled synthesis of large numbers of superparamagnetic MNPs, without a stabilizing surfactant, is the co-precipitation of the iron salts in an alkaline environment (Roth, Schwaminger, Schindler, Wagner, & Berensmeier, 2015).

The co-precipitation method is a very easy method for the synthesis of oxide nanoparticles. This method can also improve the crystal structure and material properties. Improvement of the structure and material properties can be optimized by considering parameters such as precursors, material preparation, pH, temperature, stirring speed and time, and surfactant concentration (Sasongko, Puspitasari, Sukarni, & Yazirin, 2018).

Many of the earliest nanoparticle syntheses were accomplished by co-precipitation of slightly soluble products from aqueous solutions followed by thermal decomposition of these products to oxides. Co-precipitation reactions involve simultaneous processes of nucleation, growth, hardening, and agglomeration. Due to the difficulty in isolating each process for independent study, the fundamental mechanisms of coprecipitation are still not fully understood. Chemical reactions that produce products with low solubility are generally chosen, so that the solution could quickly reach a saturated state. The chemical reactions used to trigger co-precipitation can occur in various forms (Kotnala & Shah, 2015).

Precipitation is the process by which a phase-separated solid is formed from a homogeneous solution, after supersaturation with solid deposition has been achieved. A number of related phenomena are known, which are often not clearly discriminated against. Crystallization from solution is a process, in which solids are directly obtained in the form of crystals. Crystallization usually takes place at a relatively low supersaturation, which is largely due to temperature reduction or solvent evaporation. Precipitation is often used to describe processes, in which the formation of a solid is induced by the addition of an agent that initiates a chemical reaction or reduces solubility (antisolvent). Precipitation usually involves high supersaturation, and thus an amorphous intermediate is often obtained as the first solid to form.

According to the International Union of Pure and Applied Chemistry (IUPAC) nomenclature, co-precipitation is the simultaneous deposition of normally soluble components with macrocomponents of the same solution by the formation of mixed crystals, by adsorption, occlusion, or mechanical entrapment (Gerhard, Helmut, Ferdi, & Jens, 2008). Co-precipitation is particularly suitable for producing homogeneous distributions of catalyst components, or for the preparation of precursors with certain stoichiometry, which can be easily converted to active catalysts. If the precursor for the final catalyst is a stoichiometrically determined compound of the subsequent catalyst constituents, the calcination and/or reduction step to produce the final catalyst usually results in crystallites of very small and closely related components. This has been demonstrated for several catalytic systems. Such good dispersion of the catalyst components is difficult to achieve by other means of preparation, and thus co-precipitation will remain an important technique in the preparation of solid catalysts, despite the disadvantages associated with the process. These disadvantages are higher technological demands, difficulty in keeping up with the quality of the precipitated product during precipitation, and problems in maintaining constant product quality throughout the precipitation process, if precipitation is intermittent.

Induction of precipitation of a compound does not guarantee that the product will be monodispersed and nanoparticulate. The process of nucleation and growth regulates the particle size and morphology of the product in the precipitation reaction. The time when precipitation begins, many small crystals initially form what is called nucleation, but they tend to quickly clump together to form larger, more thermo-dynamically stable particles called growth following the phenomenon, "hardening" in which the smaller particles are essentially consumed by the larger particles during the growth process. Kinetic factors compete with the thermodynamics of the system in the growth process. Factors such as reaction rate, reactant transport rate, accommo-dation, removal, and redistribution of matter compete with thermodynamic influences on particle growth. The process of growth of the deposited particles can be either limited diffusion or limited reaction. The interfacial controlled growth of small particles in solution becomes diffusion controlled once the particles exceed a critical size. The rate of reaction and transport is affected by the concentration of the reactants, temperature, pH, and the order of entry and degree of mixing of the reagents. The structure and crystallinity of the particles can be affected by the reaction rate and impurities. Factors such as saturation, nucleation and growth rate, colloid stability, recrystallization, and aging process have an effect on particle size and morphology. At low saturation, the particles are small, compact, and well-formed and their shape depends on the crystal structure and surface energy. On the other hand, at high satur-ation, large and dendritic particles are formed, while smaller particles are compacted and agglomerated, which means that saturation shows a dominant influence on sedi-ment morphology.

10.2 MATERIAL AND METHODS

10.2.1 SYNTHESIS OF ZINC FERRITE

The materials used for the synthesis included ZnO from Merck products, Fe_2O_3 from Sigma-Aldrich products, ethylene glycol solution from Merck products, and sodium hydroxide (NaOH) solution from SAP Chemicals products. The synthesis process used to obtain $ZnFe_2O_4$ powder was the co-precipitation method. The initial step taken was a precursor in the form of iron (III) oxide (Fe_2O_3) powder and 2 g of zinc oxide (ZnO) powder, each mixed with 40 ml of ethylene glycol as a solvent. The mix-ture is then stirred at a speed of 200 rpm for 48 hours so that the precursor material could dissolve completely and a homogeneous mixture could be obtained. The next process is a titration process using 5M sodium hydroxide (NaOH) solution, which is added dropwise until the solution mixture reaches the desired pH value. The solution mixture was then heated at 80 °C to form a gel to obtain $ZnFe_2O_4$ sample deposits. The sediment obtained was carried out by the washing process using 1500 ml distilled water in stages three times. The precipitate is then dried in an oven at a temperature of 100°C. The next process is the crushing process which aims to smooth the dry precipitate, and samples of $ZnFe_2O_4$ powder were obtained. The $ZnFe_2O_4$ powder sample was then carried out by the sintering process at a temperature of 1000°C for 3.5 hours to obtain the crystallinity of the material.

10.2.2 MATERIAL CHARACTERIZATION

XRD testing was used to analyze the phase and determine the average size of the crystal. The tool used is the PanAnalytical Expert Pro with Cu Kα ($\lambda = 1.54060$ Å) and a range of 10°–90° (2θ). The mean crystal size can be determined using the Debye-Scherrer Equation (Equation 10.4) as follows:

$$D = \frac{0,9\lambda}{\beta cos\theta}$$

(10.4)

where λ is the wavelength (Cu Kα), β is Full Width Half Maximum (FWHM), and θ is the diffraction angle.

SEM testing is used to identify morphology. The instrument used was Phenom with a voltage of 20 kV and a magnification of 25k and 100k times.

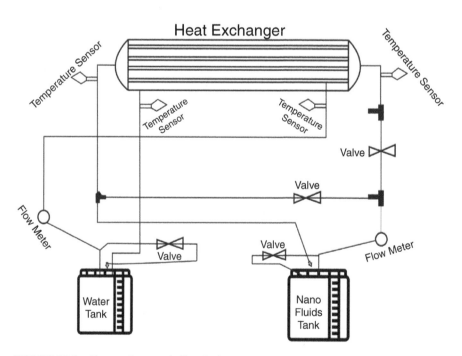

FIGURE 10.1   Heat exchanger shell and tube.

FTIR test was used to analyze the functional groups of material, and the instrument used was the Shimadzu IR Prestige21 with a wave range of 400 cm^{-1}–4000 cm^{-1}.

Heat exchanger testing was used to determine the performance of nanofluids compared to conventional fluids (distilled water). The instrument used was a shell and tube-type heat exchanger located in the Nanomaterials Laboratory of Mechanical Engineering, State University of Malang as shown at Figure 10.1. The performance of a coolant fluid in a heat exchanger test can be determined by calculating the efficiency value (η) using equation 10.5 and the logarithmic average temperature (ΔTm) using equation 10.6.

$$\eta = \frac{T_1 - T_2}{T_1 - t_1} \times 100\% \tag{10.5}$$

$$\Delta Tm = \frac{(T_1 - t_1) - (T_2 - t_2)}{\ln \dfrac{T_1 - t_1}{T_2 - t_2}} \tag{10.6}$$

where T_1 is the temperature of the hot fluid in, T_2 is the temperature of the hot fluid out, t_1 is the temperature of the cold fluid in, and t_2 is the temperature of the cold fluid out.

10.3 RESULTS AND DISCUSSION

10.3.1 X-RAY DIFFRACTION

From Figure 10.2, it can be seen that the test sample has the characteristic diffraction peaks shown at [220], [311], [222], [400], [422], [511], and [440] with each diffraction angle (2θ) 29.92o, 35.24°, 36.26°, 42.83°, 53.11°, 56.61°, and 62.16°. This shows that the structure of the $ZnFe_2O_4$ powder sample is a face-centered cubic (FCC) (Vinosha et al., 2017). There are no additional peaks in the diffraction pattern so that it can be ascertained that the samples observed are free of impurities (Sivagurunathan & Sathiyamurthy, 2016). There is no significant peak shift, which indicates that the $ZnFe_2O_4$ powder sample is a single phase, and there is no phase change in the $ZnFe_2O_4$ powder sample, which was synthesized with varying pH values (Lassoued et al., 2017).

Based on Figure 10.2 and Table 10.1, samples of $ZnFe_2O_4$ powder with varying pH values of 10 had crystals measuring 84.67 nm, while samples of $ZnFe_2O_4$ powder with variations in pH values 11 had smaller crystals, namely 38.5 nm. The decrease in crystal size was caused by the use of high concentrations of precipitating substances resulting in a shock collision between particles that occurred during the synthesis process (Kerroum et al., 2019).

Zinc ferrite powder samples with varying pH values of 12 had a larger crystal size than $ZnFe_2O_4$ powder samples with varying pH values of 11. The crystallite size of $ZnFe_2O_4$ powder using a pH value of 11 to 12 is increase. The increase in crystal size is caused by the addition of a precipitating agent or an increase in the pH value of the

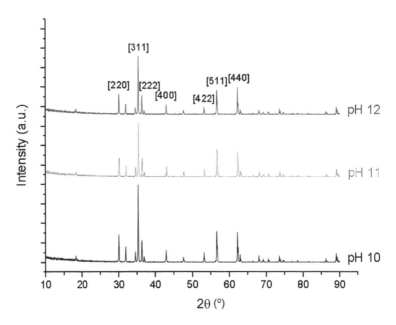

FIGURE 10.2 $ZnFe_2O_4$ diffraction pattern XRD test.

TABLE 10.1
XRD Test

Sample	Pos [°2θ]	Height [cts]	FWHM [°2θ]	d-Spacing [Å]	Crystalline [nm]
pH 10	35.24	792.07	0.0984	2.546	84.67
pH 11	35.38	557.50	0.2165	2.537	38.5
pH 12	35.27	609.08	0.0984	2.544	84.68

solution so that an acceleration of the growth phase process occurs during the synthesis process (Kerroum et al., 2019).

10.3.2   SCANNING ELECTRON MICROSCOPY

From Figure 10.3, it can be seen that the test sample tends to have a shape like an irregular octagon. Zinc ferrite powder samples with a pH value of 11 appear to have a smaller and uniform size when compared to the same sample with variations in pH values of 10 and 12. Agglomeration or buildup that occurs in samples of $ZnFe_2O_4$ powder with a variation in the pH value 11 looks higher than the variation in the pH values 10 and 12. This is because the smaller particle size has a larger surface area so that the surface energy or magnetic interaction between adjacent ions is also greater (Ramamurthi, Manimozhi, & Natesan, 2014; Patade et al., 2020). The results of the

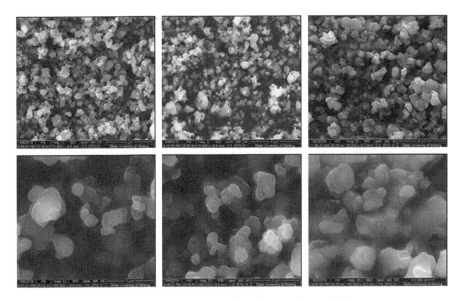

FIGURE 10.3   SEM ZnFe$_2$O$_4$ test: (a) pH 10, (b) pH 11, and (c) pH 12.

SEM test on the sample show conformity with the results of the calculation of crystal size from the results of the XRD test where the ZnFe$_2$O$_4$ powder sample with a pH value of 11 appears to have a smaller size when compared to the same sample with a pH value of 10 and 12.

10.3.3   FOURIER TRANSFORM INFRARED

Based on Figure 10.4, the distribution of functional groups of the test sample can be divided into four parts. Part I with a spectrum value of 2308–2378 cm^{-1} is a spectrum that shows the vibration stretching of the O=C=O bond (Fabbiyola et al., 2015). Part II has two parts of the spectrum value, with the first spectrum value being 1637–1654 cm^{-1}, which is a spectrum that shows vibrations due to stretching the C=C bond (Kanagesan & Ponnusamy, 2013). The second spectrum value, 1361 cm^{-1}, is a spectrum that shows vibrations due to bending the C–H bond from the methyl group (Naseri, Saion, & Kamali, 2012). Part III with a spectrum value of 867 cm^{-1} is a sign of a large change that is observed after the sample is irradiated, namely, by the emergence of vibrations of the Fe–OH bond (Rachna, Singh, & Agarwal, 2018; Sharma & Ghose, 2015). Part IV has two parts of the spectrum value, with the first spectrum value being 501 cm^{-1} and the second spectrum value being 418 cm^{-1}. This spectrum value indicates the vibration stretching of the metal-oxygen (M-O) bond contained in the test sample. This shows that the 501 cm^{-1} spectrum value confirmed the Fe–O bond at the tetrahedral position and at the 418 cm^{-1} spectrum value confirmed the Zn–O bond at the octahedral position in the ZnFe$_2$O$_4$ powder sample (Kanagesan & Ponnusamy, 2013; Shahraki, Ebrahimi, Ebrahimi, & Masoudpanah, 2012).

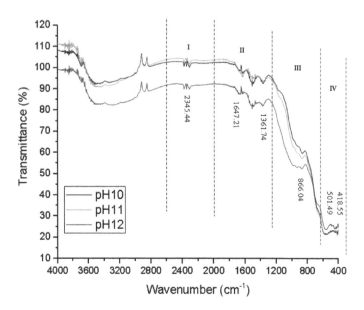

FIGURE 10.4   Molecular bonding of Zn-ferrite at different pH.

10.3.4 HEAT EXCHANGER

The results of the heat exchanger experiment on $ZnFe_2O_4$ nanofluid are shown in Table 10.2 with conventional fluid (distilled water) as a comparison. The average efficiency (η) of $ZnFe_2O_4$ nanofluid is 14.49%, while that of distilled water is 13.19%. This shows that the use of nanofluids as a cooling fluid in heat exchangers can increase the efficiency of heat transfer that occurs (Permanasari, Kuncara, et al., 2019). The average ΔTm value of $ZnFe_2O_4$ nanofluid was 28.45 °C, while that of distilled water was 27.94 °C. The high value of ΔTm is directly proportional to the amount of heat exchanged (q) based on the equation $q = U \times A \times \Delta Tm$. A large amount of heat exchanged (q) indicates the effectiveness of a heat transfer (Puspitasari, Permanasari, Shaharun, & Tsamroh, 2020). Based on previous research, the use of $ZnFe_2O_4$as a nanofluid can increase thermal conductivity when compared to conventional fluids (distilled water-ethylene glycol) (Permanasari, Puspitasari, et al., 2019; Tripathi & Chandra, 2015). Based on this, it can be ascertained that the addition of the nanomaterial as a cooling fluid mixture can be used as a nanofluid and can be used commercially in the future.

Based on research on nanofluids, it is shown that nanofluids have a higher thermal conductivity than conventional coolants and therefore it can increase the efficiency of heat transfer that occurs (Leong, Saidur, Kazi, & Mamun, 2010). The following is a discussion of the data from the test results of samples of $ZnFe_2O_4$ powder used as a nanofluid as shown in Figures 10.5 and 10.6.

The experimental results of the heat exchanger on $ZnFe_2O_4$ nanofluids are shown in Figures 5.4 and 5.5, which are compared with conventional fluids (distilled water). From Figure 5.4, it can be seen that the average efficiency (η) of $ZnFe_2O_4$ nanofluid

TABLE 10.2
Testing Results of Heat Exchanger ZnFe$_2$O$_4$ as Nanofluid

Debit	T$_1$ In [°C]	T$_2$ Out [°C]	t$_1$ In [°C]	t$_2$ Out [°C]	η [%]	ΔTm [°C]
Distilled Water						
0.2 l/min	60	55.7	27.1	32.9	13.07	27.54
0.4 l/min	60	55.7	27.2	31.5	13.1	28.28
0.6 l/min	60	55.6	27.2	31.9	13.41	28
Zinc Ferrite Nanofluid						
0.2 l/min	60	55.3	27.1	31.8	14.28	27.93
0.4 l/min	60	55.2	27.1	30.3	14.58	28.71
0.6 l/min	60	55.3	27.2	30.2	14.33	28.77

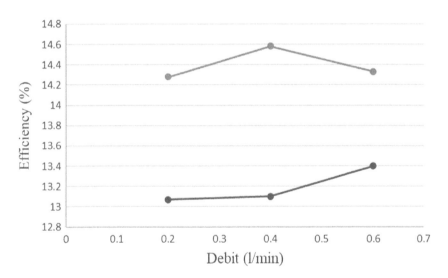

FIGURE 10.5 Experimental graph of heat exchanger efficiency.

has a value greater than the average efficiency (η) of distilled water. The average efficiency (η) of ZnFe$_2$O$_4$ nanofluid was 14.49%, while the average efficiency (η) of distilled water was 13.19%. This shows that the use of nanofluids as a cooling fluid in a heat exchanger can increase the efficiency of heat transfer that occurs.

From Figure 10.6, it can also be seen that the average value of ΔTm of ZnFe$_2$O$_4$ nanofluid has a greater value than the average value of Tm of distilled water. The average Tm value of ZnFe$_2$O$_4$ nanofluid was 28.45 °C, while the average ΔTm value of distilled water was 27.94 °C. The value of Tm can be used to calculate the amount of heat exchanged (q) using Equation 10.6. Based on this equation, it can be seen that the amount of heat exchanged (q) is directly proportional to the value of Tm, which means that the greater the value of Tm, the greater the amount of heat exchanged (q).

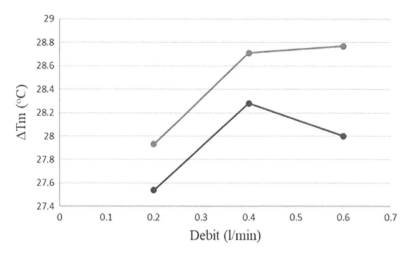

FIGURE 10.6   Experimental graph of ΔTm in heat exchanger.

The large amount of heat exchanged (q) indicates the effectiveness of a heat transfer (Puspitasari et al., 2021). Based on research that has been carried out using $ZnFe_2O_4$ as a nanofluid can increase thermal conductivity when compared to conventional fluids (distilled water-ethylene glycol) (Tripathi & Chandra, 2015). Based on this, it can be concluded that the addition of nanomaterials as a cooling fluid mixture can be used as nanofluids and can be used commercially in the future.

10.4   CONCLUSION

Zinc ferrite nanomaterial has been successfully synthesized using co-precipitation method. The treatment of variations in the value of pH 10, pH 11, and pH 12 is given in the synthesis process. The XRD test results show that the $ZnFe_2O_4$ powder sample has a single phase with an FCC structure. Based on the results of the calculation of the crystal size from the XRD test, the smallest crystal size was obtained in the $ZnFe_2O_4$ powder sample with a pH value of 11, which is 38.5 nm. SEM test results showed that samples of $ZnFe_2O_4$ powder with a pH value of 11 experienced higher agglomeration compared to samples of $ZnFe_2O_4$ with a pH value of 10 and 12. FTIR test results confirmed that the wavenumber with a spectrum value of 501 cm^{-1} confirmed the Fe–O bond at the tetrahedral position and at a spectrum value of 418 cm^{-1} it confirmed the Zn–O bond at the octahedral position in the $ZnFe_2O_4$ powder sample. The heat exchanger test results show that nanomaterials as a mixture in the cooling fluid can increase the heat transfer efficiency value of the heat transfer system.

REFERENCES

Asmin, L. O., Mutmainnah, & Suharyadi, E. (2015). Sintesis Nanopartikel Zinc Ferrite ($ZnFe_2O_4$) Dengan Metode Kopresipitasi Dan Karakterisasi Sifat Kemagnetannya. *Jurnal Fisika Dan Aplikasinya, 16*(3), 62–66.

Chen, S., Xia, J., & Dai, J. (2015). Effects of heating processing on microstructure and magnetic properties of Mn–Zn ferrites prepared via chemical co-precipitation. *Journal Wuhan University of Technology, Materials Science Edition*, *30*(4), 684–688. https://doi.org/10.1007/s11595-015-1212-8

Dar, M. A., Shah, J., Siddiqui, W. A. & Kotnala, R. K. (2014) Study of structure and magnetic properties of Ni-Zn ferrite nano-particles synthesized via co-precipitation and reverse micro-emulsion technique. *Applied Nanoscience*, *4*, 675–682. http://dx.doi.org/10.1007/s13204-013-0241-x

Fabbiyola, S., Kennedy, L. J., Ratnaji, T., Vijaya, J. J., Aruldoss, U., & Bououdina, M. (2015). Effect of Fe-doping on the structural, optical and magnetic properties of ZnO nanostructures synthesised by co-precipitation method. *Ceramics International*, *8842*(15), 1–32. https://doi.org/10.1016/j.ceramint.2015.09.110

Firlianda, D. A., Permanasari, A. A., Puspitasari, P., & Sukarni. (2019). Heat transfer enhancement using nanofluids ($MnFe_2O_4$-ethylene glycol) in mini heat exchanger shell and tube. *AIP Conference Proceedings*, *050014*(July), 1–11. https://doi.org/10.1063/1.5115690

Gajanan, K., & Tijare, S. N. (2018). Applications of nanomaterials. *Materials Today: Proceedings*, *5*(1), 1093–1096. https://doi.org/10.1016/j.matpr.2017.11.187

Gerhard, E. , Helmut, K., Ferdi, S., & Jens, W. (2008). Precipitation and coprecipitation. In *Handbook of Heterogeneous Catalysis* (2nd ed.; pp. 100–119). Weinheim: Wiley-VCH.

Holman, J. P. (2010). *Heat Transfer* (10th ed.; L. Neyens, Ed.). New York: McGraw-Hill.

Incropera, F. P., Dewitt, D. P., Bergman, T. L., & Lavine, A. S. (2007). *Fundamentals of Heat and Mass Transfer* (J. Hayton, Ed.). Los Angeles: John Willey & Sons, Inc.

Kanagesan, R. R. R. R. S., & Ponnusamy, A. K. S. (2014). Synthesis and study of structural, morphological and magnetic properties of $ZnFe_2O_4$ nanoparticles. *J Supercond Nov Magn*, *27*, 1499–1502. https://doi.org/10.1007/s10948-013-2466-z

Kandpal, N. D., Sah, N., Loshali, R., Joshi, R., & Prasad, J. (2014). Co-precipitation method of synthesis and characterization of iron oxide nanoparticles. *Journal of Scientific and Industrial Research*, *73*(2), 87–90.

Kerroum, M. A. A., Essyed, A., Iacovita, C., Baaziz, W., Ihiawakrim, D., Mounkachi, O., … Ersen, O. (2019). The effect of basic pH on the elaboration of $ZnFe_2O_4$ nanoparticles by co-precipitation method: structural, magnetic and hyperthermia characterization abstract. *Journal of Magnetism and Magnetic Materials*, *8853*(18), 1–18. https://doi.org/10.1016/j.jmmm.2019.01.081

Kotnala, R. K., & Shah, J. (2015). Ferrite materials: nano to spintronics regime. *Handbook of Magnetic Materials*, *23*, 291–379. https://doi.org/10.1016/B978-0-444-63528-0.00004-8

Lassoued, A., Ben, M., Karolak, F., Dkhli, B., Ammar, S., & Gadri, A. (2017). Synthesis and magnetic characterization of spinel ferrites—MFe_2O_4 (M = Ni, Co, Zn and Cu) via chemical co-precipitation method. *Journal of Materials Science: Materials in Electronics*, *7873*, 1–8. https://doi.org/10.1007/s10854-017-7837-y

Lee, H., Chul, J., Kim, H., Chung, Y., Jin, T., Jun, S., … Kyu, I. (2008). Effect of pH in the preparation of $ZnFe_2O_4$ for oxidative dehydrogenation of n-butene to 1,3-butadiene: correlation between catalytic performance and surface acidity of $ZnFe_2O_4$. *Catalysis Communication*, *9*, 1137–1142. https://doi.org/10.1016/j.catcom.2007.10.023

Lehyani, A., Technol, I. J. A., Sha, A. L., Ra, H., Aa, A., Alomayri, T., & Alamri, H. (2017). Magnetic hyperthermia using cobalt ferrite nanoparticles: the influence of particle size. *International Journal of Advancements in Technology*, *8*(4), 4–9. https://doi.org/10.4172/0976-4860.1000

Leong, K. Y., Saidur, R., Kazi, S. N., & Mamun, A. H. (2010). Performance investigation of an automotive car radiator operated with nano fluid-based coolants (nano fluid as a coolant in a radiator). *Applied Thermal Engineering, 30*(17–18), 2685–2692. https://doi. org/10.1016/j.applthermaleng.2010.07.019

Milanović, M., Moshopoulou, E. G., Stamopoulos, D., Devlin, E., Giannakopoulos, K. P., Kontos, A. G., … Nikolić, L. M. (2013). Structure and magnetic properties of $Zn_{1-x}In_xFe_2O_4$ and $ZnY_xFe_{2-x}O_4$ nanoparticles prepared by coprecipitation. *Ceramics International, 39*(3), 3235–3242. https://doi.org/10.1016/j.ceramint.2012.10.011

Naseri, M. G., Saion, E. B., & Kamali, A. (2012). Synthesized by a thermal treatment method. *International Scholarly Research Network, 2012*, 1–11. https://doi.org/10.5402/2012/604241

Patade, S. R., Andhare, D. D., Somvanshi, S. B., Kharat, P. B., More, S. D., & Jadhav, K. M. (2020). Preparation and characterizations of magnetic nanofluid of zinc ferrite for hyperthermia application. *Nanomaterials and Energy, 9*, 1–7. https://doi.org/10.1680/jnaen.19.00006

Permanasari, A. A., Kuncara, B. S., Puspitasari, P., Sukarni, Ginta, T. L., & Irdianto, W. (2019). Convective heat transfer characteristics of TiO_2-EG nanofluid as coolant fluid in heat exchanger. *AIP Conference Proceedings, 050015*(July), 1–12. https://doi.org/10.1063/1.5115691

Permanasari, A. A., Puspitasari, P., & Sukarni. (2019). Thermophysical properties and heat transfer performance of TiO_2—distilled water nanofluid using shell and tube heat exchanger. *International Conference on Science and Technology*, 1–6.

Puspitasari, P. (2017). Sintesis nanomaterial. *Jurnal Nanosains & Nanoteknologi ISSN, 1979*(February 2014), 0880.

Puspitasari, P., Permanasari, A. A., Shaharun, M. S., & Tsamroh, D. I. (2020). Heat transfer characteristics of NiO nanofluid in heat exchanger. *AIP Conference Proceedings, 030023*(April), 1–9. https://doi.org/doi.org/10.1063/5.0000883

Puspitasari, P., Rizkia, U. A., Sukarni, S., Permanasari, A. A., Taufiq, A., & Putra, A. B. N. R. (2021). Effects of various sintering conditions on the structural and magnetic properties of zinc ferrite ($ZnFe_2O_4$). *Materials Research, 24*(1), e20200300. https://doi.org/10.1590/1980-5373-mr-2020-0300

Qin, M., Shuai, Q., Wu, G., Zheng, B., Wang, Z., & Wu, H. (2017). Zinc ferrite composite material with controllable morphology and its applications. *Materials Science & Engineering B, 224*, 125–138. https://doi.org/10.1016/j.mseb.2017.07.016

Rachna, Singh, N. B., & Agarwal, A. (2018). Preparation, characterization, properties and applications of nano zinc ferrite. *Materials Today: Proceedings, 5*, 9148–9155. https://doi.org/10.1016/j.matpr.2017.10.035

Ramamurthi, K., Manimozhi, T., & Natesan, K. (2014). Synthesis of zinc ferrite ($ZnFe_2O_4$) nanoparticles with different capping agents. *International Journal of ChemTech Research, 7*(January), 2144–2149. https://doi.org/ISSN: 0974-4290

Roth, H. C., Schwaminger, S. P., Schindler, M., Wagner, F. E., & Berensmeier, S. (2015). Influencing factors in the CO-precipitation process of superparamagnetic iron oxide nano particles: a model based study. *Journal of Magnetism and Magnetic Materials, 377*(March), 81–89. https://doi.org/10.1016/j.jmmm.2014.10.074

Sasongko, M. I. N., Puspitasari, P., Sukarni, & Yazirin, C. (2018). Properties of MnO doped graphene synthesized by co-precipitation method. *Functional Materials, 25*(4), 802–808. https://doi.org/10.15407/fm25.04.802

Shahraki, R. R., Ebrahimi, M., Ebrahimi, S. A. S., & Masoudpanah, S. M. (2012). Structural characterization and magnetic properties of superparamagnetic zinc ferrite nanoparticles synthesized by the coprecipitation method. *Journal of Magnetism and Magnetic Materials, 324*(22), 3762–3765. https://doi.org/10.1016/j.jmmm.2012.06.020

Sharma, R. K., & Ghose, R. (2015). Synthesis and characterization of nanocrystalline zinc ferrite spinel powders by homogeneous precipitation method. *Ceramics International, 8842*(15). 14684–14691. https://doi.org/10.1016/j.ceramint.2015.07.191

Sivagurunathan, P., & Sathiyamurthy, K. (2016). Effect of temperatures on structural, morphological and magnetic properties of zinc ferrite. *Nanoparticles, 4*(2), 244–254. https://doi.org/10.13179/canchemtrans.2016.04.02.0299

Tripathi, A., & Chandra, H. (2015). Performance investigation of automobile radiator operated with $ZnFe_2O_4$ nano fluid based coolant. *EDP Sciences, 3*, 1–7. https://doi.org/doi.org/10.1051/matecconf/20153401003

Vinosha, P. A., Mely, L. A., Jeronsia, J. E., Krishnan, S., & Das, S. J. (2017). Synthesis and properties of spinel $ZnFe_2O_4$ nanoparticles by facile co-precipitation route. *Optik - International Journal for Light and Electron Optics, 4026*(17), 1–22. https://doi.org/10.1016/j.ijleo.2017.01.018

11 A Study of Risk Assessment in the Nanomaterials Laboratory of Mechanical Engineering Department and the Materials Physics Laboratory of Department of Physics at State University of Malang

Djoko Kustono,[1,*] *Desi Puspita Anggraeni,*[1]
and Poppy Puspitasari[1,2]
[1] Faculty of Engineering, Universitas Negeri Malang, Indonesia
[2] Centre of Advanced Materials for Renewable Energy (CAMRY), State University of Malang, Indonesia
[*] Corresponding author: Djoko Kustono, djoko.kustono.ft@um.ac.id

CONTENTS

DOI: 10.1201/9781003320746-11

11.1 INTRODUCTION

Nanotechnology is a new science that has been widely applied to daily life products, which marks technological advances in the Industrial Era 4.0. In line with the development of technology, experts in the fields of nanotechnologies, biotechnologies, robotics, and chemistry will be increasingly needed. The State University of Malang has prepared its graduates, especially graduates majoring in Mechanical Engineering and Physics, through the provision of an educational curriculum that focuses on nanotechnology so that graduates can compete in and adapt to the world of work in the Industrial Era 4.0.

One of the subjects taught in the Material Expertise of the Mechanical Engineering Department and the Material Expertise of the Material Physics Department is the synthesis and fabrication of materials. Synthesis of nanomaterials is an activity that has the potential for exposure to nanomaterials entering the body, because the nanomaterial size is very small, namely 1 to 100 nm. Nanomaterials can enter the human body through the respiratory, digestive, and skin contact pathways [1]. Nano-sized particles or nanomaterials (1–100 nm) have greater potential than particles <100 nm in size in triggering lung inflammation because nanomaterials are more easily retained in the alveolar area [2].

To prevent and reduce the exposure of nanomaterials from entering the body, personal protective equipment (PPE) is used when carrying out nanomaterials synthesis. The types of PPE used in the nanomaterial laboratory are based on the nanomaterial laboratory level. This level is obtained through a decision tree analysis of the level of potential hazards in the nanomaterial laboratory, and then the results are analyzed to obtain a nanomaterial laboratory level [3]. In addition to using PPE, nanomaterial synthesis practitioners need to know how to properly handle nanomaterials to prevent exposure to nanomaterials.

After making these efforts, it is necessary to carry out a risk assessment to raise awareness of the hazards and risks and to determine risk control measures. Risk control in the nanomaterial laboratory in the Department of Mechanical Engineering and the Department of Physics can be carried out based on a hierarchy of control which consists of five steps, namely, elimination, substitution, engineering controls, administrative controls, and PPE [4], [5].

11.2 METHODOLOGY

This research is a descriptive study using a qualitative approach. The type of research used in this research is a multi-case study because this research was conducted to uncover events regarding risk assessment in the nanomaterial laboratory of the Department of Mechanical Engineering and the Department of Materials Physics at State University of Malang. The selected sources to obtain information about the application of nanosafety in the nanomaterial laboratory are the following:

a. Lecturer of the Materials Expertise of Mechanical Engineering Department
b. Nanomaterial Laboratory Assistant of Mechanical Engineering Department
c. Students of the Materials Expertise of Mechanical Engineering Department
d. Lecturer of the Materials Physics Expertise of Physics Department
e. Nanomaterials Laboratory Assistant of Physics Department
f. Students of the Materials Physics of Physics Department

Data collection techniques used in this study included interviews, observation, and documentation. Qualitative data analysis has three stages, namely, data reduction, data display, and drawing conclusions and verification [6]. The validity of the results of this study can be tested using the triangulation method. There are four types of triangulation, namely, source triangulation, method triangulation, investigator triangulation, and theoretical triangulation [7]. The triangulation methods used in this research are source triangulation and method triangulation. Source triangulation is done by collecting the results of interviews from several sources, while method triangulation is done by comparing the data from interviews and observations.

11.3 RESULTS

11.3.1 THE LABORATORY USERS' KNOWLEDGE OF NANOSAFETY

Based on the results of interviews and observations that have been carried out, it can be concluded that the level of knowledge of laboratory users on nanosafety in the Department of Mechanical Engineering is in the moderate/sufficient category, while the level of knowledge of laboratory users on nanosafety in the Department of Physics is in the high category.

11.3.2 The Condition of Nanosafety Facilities in Nanomaterials Laboratory

Based on the results of interviews and observations that have been carried out, it was found that the conditions of the nanosafety facilities in the two laboratories were in the incomplete category. The results of observations at the Nanomaterials Laboratory of the Department of Mechanical Engineering are presented in Table 11.1, and the results of observations at the Nanomaterials Laboratory of the Department of Physics are presented in Table 11.2.

TABLE 11.1
Observation Results of Nanosafety Facilities in the Nanomaterials Laboratory of Mechanical Engineering Department

Observation Sheet
Location: Nanomaterials Laboratory of Mechanical Engineering Department, State University of Malang
Day, date: Thursday, 19 December 2019

No.	Aspects of Observation	Yes/ Available/ Adequate	No/ Not Available/ Inadequate	Explanation
A. PPE [3]				
1st Level Nanomaterials Laboratory				
1.	Safety glasses			
2.	The amount of safety glasses suits the number of laboratory users			
3.	Simple lab coat			
4.	The amount of simple lab coats suits the number of laboratory users			
5.	One pair of gloves			
6.	The amount of gloves suits the number of laboratory users			
2nd Level Nanomaterials Laboratory				
1.	Safety glasses	Available		
2.	The amount of safety glasses suits the number of laboratory users	Yes/ Adequate		Provided according to the maximum number of researchers in the room
3.	Non-woven lab coat	Available		
4.	The amount of non-woven lab coats suits the number of laboratory users	Yes/ Adequate		Provided according to the maximum number of researchers in the room
5.	Overshoes		Not Available	

TABLE 11.1 (Continued)
Observation Results of Nanosafety Facilities in the Nanomaterials Laboratory of Mechanical Engineering Department

6.	The amount of overshoes suits the number of laboratory users	Not Available	
7.	Two pairs of gloves	Available	
8.	The amount of gloves suits the number of laboratory users	Yes/ Adequate	There is a lot of stock

3rd Level Nanomaterials Laboratory

1. Safety goggles
2. The amount of safety goggles suits the number of laboratory users
3. Mask with ventilation aid (for work of 2 hours or more)
4. The amount of mask with ventilation aid suits the number of laboratory users
5. FFP3 mask (for work less than 2 hours)
6. The amount of FFP3 mask suits the number of laboratory users
7. Overall with hood
8. The amount of overall suits the number of laboratory users
9. Overshoes
10. The amount of overshoes suits the number of laboratory users
11. Two pairs of gloves
12. The amount of gloves suits the number of laboratory users

B. Nanosafety, Hygiene, and Spill Kit Equipment

1.	Fume hoods/ Biological Safety Cabinet/ Glovebox	Available	Provided with fume hood
2.	Vacuum Cleaner with HEPA filter (for large-scale nanomaterial spills)	Not Available	Spilled nanomaterial cases are extremely rare, and if they did, the scale is small
3.	Sealed plastic bag as a trash bag	Not Available	Nanomaterial wastes are disposed of in the usual trash
4.	Bench Liner	Not Available	
5.	Barricade tape (for large-scale nanomaterial spills)	Not Available	Spilled nanomaterial cases are extremely rare, and if they did, the scale is small

(continued)

TABLE 11.1 (Continued)
Observation Results of Nanosafety Facilities in the Nanomaterials Laboratory of Mechanical Engineering Department

6.	Nitrile gloves	Available		There is a lot of stock
7.	Mask (respiratory protection)	Available		There is a lot of stock
8.	Walk-off mat		Not Available	
9.	Absorbent cloth		Not Available	Wipe using a tissue
10.	Sink	Available		
11.	Fire extinguisher	Available		
12.	Mop (for wet cleaning)	Available		

C. Material Information

1.	The material is stored in closed containers	Yes		Stored in a vial or bottle and put in a dry box
2.	Containers are labeled with the name of the material	Yes		
3.	MSDS		Not Available	
4.	Labels of flammable, explosive, etc.	Yes		Available on the packaging

D. Standard Operating Procedures and Emergency Procedures Information

1.	SOP is installed in the work area	No
2.	Emergency procedures are installed in the work area	No

TABLE 11.2
Observation Results of Nanosafety Facilities in the Nanomaterials Laboratory of Physics Department

Observation Sheet
Location: Nanomaterials Laboratory of Physics Department, State University of Malang
Day, date: Thursday, 5 December 2019

No.	Aspects of Observation	Yes/ Available/ Adequate	No/ Not Available/ Inadequate	Explanation
A. PPE [3]				
1st Level Nanomaterials Laboratory				
1.	Safety glasses			
2.	The amount of safety glasses suits the number of laboratory users			
3.	Simple lab coat			
4.	The amount of simple lab coats suits the number of laboratory users			
5.	One pair of gloves			
6.	The amount of gloves suits the number of laboratory users			

TABLE 11.2 (Continued)
Observation Results of Nanosafety Facilities in the Nanomaterials Laboratory of Physics Department

2nd Level Nanomaterials Laboratory

1.	Safety glasses	Available	
2.	The amount of safety glasses suits the number of laboratory users	Yes/ Adequate	Provided according to the maximum number of researchers in the room
3.	Non-woven lab coat	Available	
4.	The amount of non-woven lab coats suits the number of laboratory users	Yes/ Adequate	Provided according to the maximum number of researchers in the room
5.	Overshoes	Available	
6.	The amount of overshoes suits the number of laboratory users	Yes/ Adequate	There is a lot of stock
7.	Two pairs of gloves	Available	
8.	The amount of gloves suits the number of laboratory users	Yes/ Adequate	There is a lot of stock

3rd Level Nanomaterials Laboratory

1. Safety goggles
2. The amount of safety goggles suits the number of laboratory users
3. Mask with ventilation aid (for work of 2 hours or more)
4. The amount of mask with ventilation aid suits the number of laboratory users
5. FFP3 mask (for work less than 2 hours)
6. The amount of FFP3 mask suits the number of laboratory users
7. Overall with hood
8. The amount of overall suits the number of laboratory users
9. Overshoes
10. The amount of overshoes suits the number of laboratory users
11. Two pairs of gloves
12. The amount of gloves suits the number of laboratory users

B. Nanosafety, Hygiene, and Spill Kit Equipments

1.	Fume hoods/ Biological Safety Cabinet/ Glovebox	Available	Provided with fumehood
2.	Vacuum cleaner with HEPA filter (for large-scale nanomaterial spills)	Not Available	Spilled nanomaterial cases are extremely rare, and if they did, the scale is small

(continued)

TABLE 11.2 (Continued)
Observation Results of Nanosafety Facilities in the Nanomaterials Laboratory of Physics Department

3.	Sealed plastic bag as a trash bag	Available		
4.	Bench Liner		Not Available	
5.	Barricade tape (for large-scale nanomaterial spills)		Not Available	Spilled nanomaterial cases are extremely rare, and if they did, the scale is small
6.	Nitrile gloves	Available		There is a lot of stock
7.	Mask (respiratory protection)	Available		There is a lot of stock
8.	Walk-off mat		Not Available	
9.	Absorbent cloth	Available		In the form of a chamois cloth
10.	Sink	Available		
11.	Fire extinguisher	Available		
12.	Mop (for wet cleaning)	Available		
C. Material Information				
1.	The material is stored in closed containers	Yes		Stored in a vial or bottle
2.	Containers are labeled with the name of the material	Yes		
3.	MSDS	Yes		Obtained online
4.	Labels of flammable, explosive, etc.	Yes		Available on packaging
D. Standard Operating Procedures and Emergency Procedures Information				
1.	SOP is installed in the work area	Yes		
2.	Emergency procedures are installed in the work area		No	Introduced at the beginning of the lesson

11.3.3 THE ACTIVITIES IN THE NANOMATERIAL LABORATORY

In the Nanomaterials Laboratory of the Mechanical Engineering Department, activities that have the potential to produce exposure to nanomaterials, such as the synthesis of nanomaterials, are carried out in 2–4 hours/8-hour shifts. Meanwhile, at the Nanomaterials Laboratory of the Department of Physics, synthesis activities are carried out in 1–2 hours/8-hour shifts. Exposure rating based on the duration of work is divided into five levels, where level 1 is the lowest level of exposure and level 5 is the highest level of exposure [8], as explained below.

Level 1 = < 1 hour/8-hour shifts
Level 2 = 1–2 hours/8-hour shifts
Level 3 = 2–4 hours/8-hour shifts
Level 4 = 4–7 hours/8-hour shifts
Level 5 = > 7 hours/8-hour shifts

Based on this explanation, the results of the activity level in the nanomaterial laboratory of the Mechanical Engineering Department are at level 3, while those in the Physics Department are at level 2.

11.3.4 Risk Assessment in Nanomaterials Laboratory

Formula to calculate the possible risks caused by nanomaterials or chemicals are as follows [9], [10]:

$$Risk = Hazard \times Exposure$$

Meanwhile, to measure the risk rating or risk level, you can use the following formula [11]:

$$RR = \sqrt{(HR \times ER)}$$

Hazard rating (HR) is obtained from the condition of the nanosafety facility in the nanomaterial laboratory and the level of knowledge of the nanomaterial laboratory users regarding the nanosafety of the nanomaterial laboratory. HR is divided into five levels, where level 1 is the lowest hazard level and level 5 is the highest hazard level, as explained in Table 11.3.

The exposure rating (ER) is obtained from the duration of work of the nanomaterial laboratory user. ER based on the duration of work is divided into five levels, where level 1 is the lowest level of exposure and level 5 is the highest level of exposure [8].

Based on the results of interviews and observations that have been made, the following results are obtained.

11.3.5 Risk Rating of Nanomaterials Laboratory of Mechanical Engineering Department

(HR is classified into fourth level, namely, the nanosafety facilities in the laboratory are incomplete and the laboratory users' knowledge of nanosafety is classified as moderate. As for the ER, it is the third level, with a working duration of 2–4 hours/8-hour shift. Based on the results of HR and ER, the risk rating matrix obtained is RR = 12 or moderate risk, as presented in Table 11.4.

TABLE 11.3
Hazard Rating of the Nanomaterials Laboratory of the Mechanical Engineering Department and the Physics Department

Hazard Rating	Nanosafety Facilities Condition	Level of Knowledge about Nanosafety
1	Complete	High
2	Complete	Moderate
3	Incomplete	High
4	Incomplete	Moderate
5	Incomplete	Low

TABLE 11.4
Risk Rating Matrix of Nanomaterials Laboratory in the Department of Mechanical Engineering

			Exposure Rating (ER)				
			1	2	3	4	5
Hazard Rating (HR)		1	RR = 1	RR = 2	RR = 3	RR = 4	RR = 5
		2	RR = 2	RR = 4	RR = 6	RR = 8	RR = 10
		3	RR = 3	RR = 6	RR = 9	RR = 12	RR = 15
		4	RR = 4	RR = 8	RR = 12	RR = 16	RR = 20
		5	RR = 5	RR = 10	RR = 15	RR = 20	RR = 25

Color Description:

    = Low Risk (RR=1 s.d. RR=4)

    = Moderate Risk (RR=5 s.d. RR=12)

    = High Risk (RR=15 s.d. RR=25)

TABLE 11.5
Risk Rating Matrix of Nanomaterials Laboratory in the Department of Physics

			Exposure Rating (ER)				
			1	2	3	4	5
Hazard Rating (HR)		1	RR = 1	RR = 2	RR = 3	RR = 4	RR = 5
		2	RR = 2	RR = 4	RR = 6	RR = 8	RR = 10
		3	RR = 3	RR = 6	RR = 9	RR = 12	RR = 15
		4	RR = 4	RR = 8	RR = 12	RR = 16	RR = 20
		5	RR = 5	RR = 10	RR = 15	RR = 20	RR = 25

Color Description:

    = Low Risk (RR=1 s.d. RR=4)

    = Moderate Risk (RR=5 s.d. RR=12)

    = High Risk (RR=15 s.d. RR=25)

11.3.6 RISK RATING OF NANOMATERIALS LABORATORY OF PHYSICS DEPARTMENT

Hazard rating (HR) is classified into third level, namely, the nanosafety facilities in the laboratory are incomplete and the laboratory users' knowledge of nanosafety is classified as high. As for the ER, it is the second level, with a working duration of 1–2 hours/8-hour shift. Based on the results of HR and ER, the risk rating matrix obtained is RR = 6 or moderate risk, as presented in Table 11.5.

11.4   DISCUSSION

11.4.1   The Laboratory Users' Knowledge of Nanosafety

The nanomaterials laboratory users' knowledge in the Department of Mechanical Engineering regarding nanosafety is in the moderate/adequate category, while in the Department of Physics the laboratory users' knowledge of nanosafety is high. This difference is because laboratory users in the Mechanical Engineering Department do not understand Material Safety Data Sheet (MSDS). When laboratory users prepare the materials for nanomaterials synthesis, an MSDS is needed to provide the users the information regarding the chemical and physical hazards of the materials used, as well as information about how to use, handle, and store the materials safely [12]. MSDS can be obtained when laboratory users purchase materials through online searches. Because users of the Nanomaterials Laboratory of the Department of Mechanical Engineering have not been optimal in the application of MSDS, knowledge of information on the toxicity and nanosafety can be stated as lacking. In addition to the lack of knowledge about MSDS, users of the Nanomaterial Laboratory of the Mechanical Engineering Department are also not optimal in the aspects of proper PPE use, knowledge of procedures for cleaning nanomaterial laboratories, and procedures for handling nanomaterial waste. This deficiency is caused by the absence of introductory lessons of nanosafety given to laboratory users before carrying out practicum activities or before studying nanomaterial synthesis.

11.4.2   The Condition of Nanosafety Facilities in Nanomaterials Laboratory

The conditions of nanosafety facilities in the Nanomaterial Laboratory of the Mechanical Engineering Department and the Physics Department are in the incomplete category. This is because the Mechanical Engineering Department's Nanomaterial Laboratory has not provided a complete PPE, which is not providing the overshoes that are used as foot protectors. In addition, there are no sealed plastic bags, bench liners, walk-off mats, and absorbent cloths as cleaning equipment and spill kits. In the Nanomaterial Laboratory, there is also no Standard Operational Procedure (SOP) sheet and emergency procedure sheet. Meanwhile, in the Nanomaterial Laboratory of the Physics Department, the PPE is complete, it's just that there are still no bench liners and walk-off mats as spill kits, and there is still no emergency procedure sheet in the laboratory.

11.4.3   The Activity in the Nanomaterials Laboratory

Several activities have the potential to be exposed to nanomaterials, namely (1) receiving nanomaterials, unpacking, and sending nanomaterials; (2) activities in the laboratory; (3) cleaning and maintaining laboratory and practicum equipment; (4) storage, packaging, and shipping; and (5) nanomaterial waste management [13]. One of the activities with the highest potential for exposure to nanomaterials is activities in the laboratory that are directly related to nanomaterials, such as synthesis, analysis, and quality assurance activities. In this study, the activity observed was nanomaterial synthesis. The synthesis process carried out in the Nanomaterial Laboratory of the Department of Mechanical Engineering is performed in 2–4 hours/ 8-hour shifts and the exposure rating is at level 3, whereas in the Physics Department, nanomaterial synthesis activities are carried out in 1–2 hours/8 hours shift with an exposure rating at level 2. According to the Department of Occupational Safety and Health (2018), the duration of work in which Nanomaterial Laboratory users are exposed to nanomaterials (exposure duration) is directly proportional to the number of nanomaterial exposures that occur, so if the duration of exposure become twice higher, it will produce the exposure level in the work environment at the same high.

11.4.4   RISK ASSESSMENT IN NANOMATERIALS LABORATORY

11.4.4.1   Risk Assessment in Nanomaterials Laboratory of Mechanical Engineering Department

The risk rating obtained is RR = 12, where the hazard rating is at level 4 and the exposure rating is at level 3. The risks in the laboratory are significant, so a strategy is needed to reduce the risk rating to risk level 1 or 2 (insignificant risk). The strategy that can be followed to reduce the level of risk is to reduce the level of hazard that was originally at level 4 (nanosafety facilities in the laboratory are incomplete and laboratory users' knowledge of nanosafety is moderate) to level 1 hazard, where nanosafety facilities in the laboratory are complete and laboratory users' knowledge of nanosafety is high. Based on this explanation, several ways can be followed to reach the hazard rating level 1, which are as follows:

- Completing the PPE that must be used in a level 2 nanomaterial laboratory (safety glasses, overshoes, lab coat, and two pairs of gloves).
- Equip cleaning equipment and supplies in the nanomaterial laboratory.
- Provide a spill kit to reduce the exposure to the nanomaterial when the nanomaterial is spilled.
- Provide information on SOPs and emergency procedures at work.
- Increase the knowledge and understanding of laboratory users on the MSDS of the materials to be synthesized.
- Increase laboratory users' understanding of laboratory cleaning procedures and handling of nanomaterial waste.
- Improve the discipline of using PPE when doing synthesis.

11.4.4.2   Risk Assessment in Nanomaterials Laboratory of Physics Department

The risk rating obtained is RR = 6, where the HR is at level 3 and the ER is at level 2. The risks in the laboratory are significant, so a strategy is needed to reduce the risk level to risk level 1 or 2 (insignificant risk). The strategy that can be followed to reduce the level of risk is to reduce the level of hazard that was originally at level 3 (nanosafety facilities in the laboratory are incomplete and laboratory users' knowledge of nanosafety is high) to level 1 hazard, where nanosafety facilities in the laboratory are complete and laboratory users' knowledge of nanosafety is high. Based on this explanation, several ways can be followed to reach the level of danger level 1, which are as follows:

- Provide a spill kit to reduce exposure to the nanomaterial when the nanomaterial is spilled.
- Provide information on emergency procedures at work.

11.4.5 Risk Control in Nanomaterials Laboratory

Risk control is a part of risk management, where risk control provides methods or measures to reduce the risks/hazards that have been identified and assessed. The main approach in controlling risk is called the hierarchy of control or the STOP principle (substitution including elimination, technical or engineering control, organizational or administrative control, and PPE) [14]. In line with Berges' statement, Morris and Cannady, and Engeman et al., divided the hierarchy of control into five steps, namely elimination, substitution, engineering controls, administrative controls, and PPE (Engeman et al., 2013; Morris & Cannady, 2019). The hierarchy of control is depicted in Figure 11.1, which shows the most effective to the least effective steps in controlling risk.

11.4.5.1 Elimination

This step is the most effective in minimizing or eliminating the risk of hazards because it directly eliminates or avoids the use of hazardous materials and processes that can

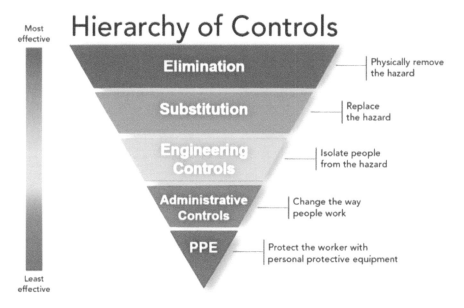

FIGURE 11.1 Hierarchy of control.

cause exposure to nanomaterials. Although this step is classified as the most effective, researchers often cannot eliminate certain materials because these materials are the core material of their research. However, changing the work process of nanomaterials from dry powder to nanomaterials suspended in liquid can reduce the potential of nanomaterials to be exposed [13].

11.4.5.2 Substitution

Substitution can be done in several ways as follows [14]:

- Replacing raw/pure hazardous materials with materials of a lower hazard/toxicity level.
- Changing the physical form of the material from dry powder or aerosol to a paste, granule, or composite.

One example of risk control with a substitution step is replacing nitric acid with citric acid and distilled water as a solvent, as has been done in the Nanomaterial Laboratory of the Department of Mechanical Engineering.

11.4.5.3 Engineering Controls

This step is an effort to reduce contamination of workers by controlling hazards or providing a barrier between workers and hazards. Engineering control is an effort to isolate the process of working with nanomaterials or equipment or contain the hazard [15]. In Figure 11.2, the factors affecting the selection of engineering controls are presented. From Figure 11.2, it is cleared that the greater the risk of exposure to nanomaterials, the more closed engineering controls are required. According to the research results where the Nanomaterial Laboratory of the Department of Mechanical

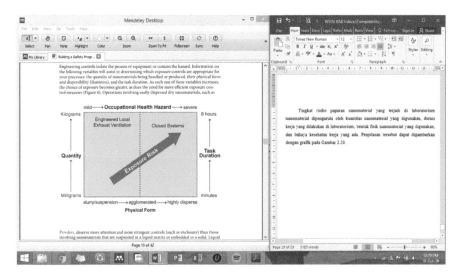

FIGURE 11.2   Factors affecting the selection of engineering controls.

Engineering and the Department of Physics were at a moderate level of risk, the engineering controls carried out included the use of a fume hood in the laboratory.

11.4.5.4 Administrative Controls

Administrative control can be carried out by reducing or limiting the duration of work that is potentially exposed to nanomaterials, establishing good storage and work procedures, providing training to workers (laboratory users), and implementing proper nanomaterial labeling and storage. Handwashing facilities should always be used before eating, drinking, smoking, or leaving the laboratory [16]. Food and beverages are not allowed in areas where nanomaterials are being processed. Walk-off mat or tacky mats are recommended to be installed to reduce dust in the workplace area and prevent tracking of materials from production areas to non-production areas [16]. Environment, Safety, and Health Division of Stanford University describes several administrative control activities as follows [17]:

- Clean the nanomaterial using a wet cloth.
- Transferring nanomaterials between workplaces (such as fume hoods, glove boxes, and furnaces) in sealed and labeled containers.
- Label it with a clear indication that the material is a nanomaterial, for example, NANOSCALE ZINC OXIDE PARTICLES or the like, and not written as ZINC OXIDE only.
- Label or mark the nanomaterials being processed or synthesized at the planned location (such as in the fume hood, glove box, or furnace) as shown in Figure 11.3.

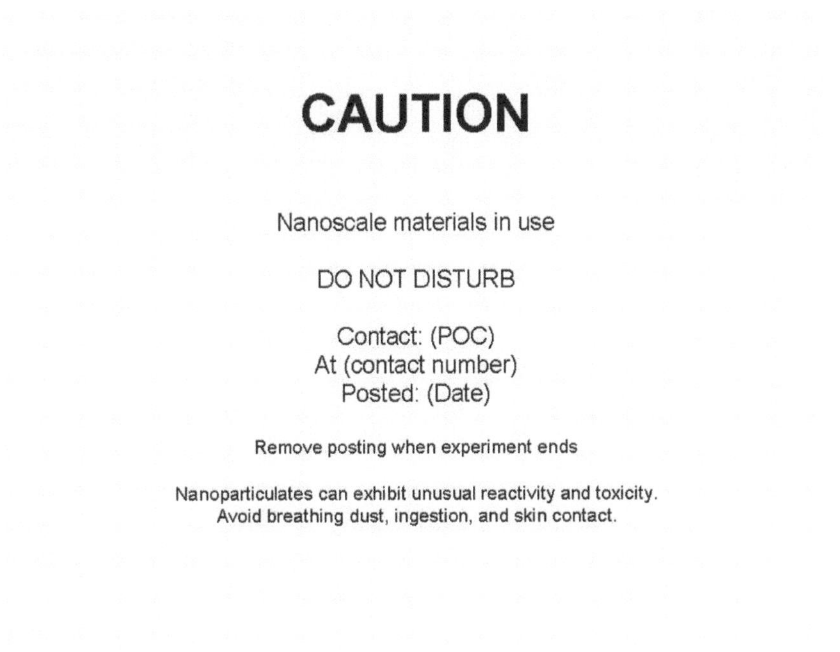

CAUTION

Nanoscale materials in use

DO NOT DISTURB

Contact: (POC)
At (contact number)
Posted: (Date)

Remove posting when experiment ends

Nanoparticulates can exhibit unusual reactivity and toxicity.
Avoid breathing dust, ingestion, and skin contact.

FIGURE 11.3   Sign of nanomaterial in processing.

Administrative controls also include cleaning and maintenance procedures for the nanomaterial laboratory and training of nanomaterial laboratory users as described below [18].

- Clean work areas and equipment every day.
- The standard approach to cleaning up nanomaterial spills is to use a wet cloth to wipe dry powder, or to wet the nanomaterial powder before wiping.
- Dispose of cleaning cloths and other contaminated material in sealed plastic bags. The liquid nanomaterial waste is isolated from ordinary water channels.
- Do not use brooms, brushes, or other dry methods.
- Provide nanosafety training to laboratory users, including information on proper use and maintenance.
- Provide laboratory users with sufficient information to understand potential occupational exposures, health risks, routes of exposure to the body, and instructions for reporting health symptoms.
- Ensure that training includes how to keep exposure to a low level, how to use and care for PPE, how to make sure and use a fume hood, and what to do if something goes wrong.

Another point in carrying administrative control is to provide relevant information regarding operating instructions and procedures when handling spilled nanomaterials [19], [14]. SOP for synthesis equipment needs to be provided in the nanomaterial laboratory.

11.4.5.5 PPE

PPE is the last resort in controlling risk because it cannot completely reduce the risk of exposure to nanomaterials. The PPE used are safety glasses, lab coat, two pairs of gloves, and overshoes, according to the nanomaterial laboratory level of the Department of Mechanical Engineering and the Department of Physics, namely, the level 2 nanomaterial laboratory. Additional informations regarding the use of PPE are presented below [17].

- Wear nitrile rubber gloves or other polymer gloves when handling nanomaterials.
- Keep gloves that are used or contaminated in a sealed plastic bag or other sealed container.
- Always wash hands and forearms after wearing gloves.
- PPE and clothing that are contaminated are to be kept separated in the laboratory or change area so that nanomaterials cannot be transported to common areas.

Based on the five steps of the hierarchy of controls described above, a conclusion can be drawn that describes the risk control that has been carried out by the Nanomaterial Laboratory in the Department of Mechanical Engineering and the Department of Physics as follows:

- The process of eliminating the risk of hazards cannot be done optimally because the nanomaterial used is the core material of the research.

- A substitution process has been carried out, demonstrated by changing a hazardous solvent to a solvent with a lower hazard level.
- Engineering controls have been carried out by both laboratories by providing a fume hood in the laboratory.
- The administrative control process still needs to be improved by following several procedures described by national Institute for Occupational Safety and Health (NIOSH), Standford Linear Accelarator Center (SLAC) National Accelerator Laboratory, and Berges et al.
- The last process in the form of PPE has been carried out, but the Nanomaterial Laboratory of the Mechanical Engineering Department needs to add overshoes per PPE standards at level 2 nanomaterial laboratories.

As risk assessment is carried out in the two nanomaterials laboratories and the outcomes state that there are things that need to be improved, and the responsibility to carry out this improvement task can be handled by several authorities such as: .

- The manager of the Nanomaterial Laboratory at the Department of Mechanical Engineering and the Department of Physics, State University of Malang

Laboratory managers are responsible to carry out administrative control based on several procedures provided by NIOSH, SLAC National Accelerator Laboratory, and Berges et al. It is intended that activities in the nanomaterial laboratory are carried out in an orderly, safe, and controlled manner. The risk of exposure to nanomaterials can also be reduced so that the negative impacts of the nanomaterials can be prevented. In addition, it is recommended to complete nanosafety facilities in the nanomaterial laboratory, such as PPE, cleaning equipment, and spill kit equipment.

- The Lecturers of Nanomaterials Subject

The lecturers are responsible in providing an introduction to nanosafety in the nanomaterial laboratory for laboratory users and students, so that before carrying out nanomaterial synthesis or other activities that require laboratory use, laboratory users and students already have an understanding of the proper procedures for handling nanomaterials and the possible risks it possessed. Nanomaterials lecturers are also advised to familiarize students with reading and understanding the MSDS of the material to be synthesized because in the MSDS there is important information regarding the handling of the material to be used.

- The Students Using Nanomaterials Laboratory

For the students who are using nanomaterial laboratories, they are responsible to add insight into the nanosafety of the nanomaterial laboratory through explanations from lecturers, research journals, and nanosafety procedures for nanomaterial laboratories from other universities. It is intended that students have extensive knowledge about the nanosafety of nanomaterial laboratories, and have the awareness to protect themselves and colleagues from exposure to nanomaterials. Students are also advised to always get used to reading MSDS before synthesizing nanomaterials so that they could know important information about the material to be synthesized.

11.6 CONCLUSIONS

In this chapter, the risk rating found at the Nanomaterials Laboratory of the Mechanical Engineering Department is 12 or moderate risk. Meanwhile, at the Nanomaterials Laboratory of Physics Department it is 6 or moderate risk.

Risk control can be carried out using the principle of a hierarchy of controls consisting of elimination, substitution, engineering control, administrative control, and PPE. Based on research that has been conducted, few points can be concluded regarding risk control in these two laboratories: (1) a substitution process has been carried out, demonstrated by changing a hazardous solvent to a solvent with a lower hazard level;, (2) engineering controls have been carried out by both laboratories by providing a fume hood in the laboratory; (3) the administrative control process still needs to be improved by following several procedures described by NIOSH, SLAC National Accelerator Laboratory, and Berges et al.; and (4) the last process in the form of PPE has been carried out, but the Nanomaterial Laboratory of Mechanical Engineering Department needs to add overshoes per PPE standards at level 2 nanomaterial laboratories.

ACKNOWLEDGMENTS

The authors would like to thank the informants for the help in interview session and the manager of Nanomaterials Laboratory at the Department of Mechanical Engineering and the Department of Physics at State University of Malang for the help during observation process.

REFERENCES

[1] B. Zhang, I. Ahmed, P. Wang, and Y. He, "Nanomaterials in the Environment and Their Health Effects," in *Encyclopedia of Environmental Health*, Elsevier, 2019, pp. 535–540. doi: 10.1016/B978-0-12-409548-9.11057-7.

[2] H. Alenius, J. Catalán, H. Lindberg, H. Norppa, J. Palomäki, and K. Savolainen, "Nanomaterials and Human Health," in *Handbook of Nanosafety*, Elsevier, 2014, pp. 59–133. doi: 10.1016/B978-0-12-416604-2.00003-2.

[3] A. Groso, A. Petri-Fink, B. Rothen-Rutishauser, H. Hofmann, and T. Meyer, "Engineered Nanomaterials: Toward Effective Safety Management in Research Laboratories," *J Nanobiotechnol*, vol. 14, no. 1, p. 21, Dec. 2016, doi: 10.1186/s12951-016-0169-x.

[4] G. A. Morris and R. Cannady, "Proper Use of the Hierarchy of Controls," *Professional Safety*, vol. 64, no. 08, pp. 37–40, Aug. 2019.

[5] C. D. Engeman *et al.*, "The Hierarchy of Environmental Health and Safety Practices in the U.S. Nanotechnology Workplace," *Journal of Occupational and Environmental Hygiene*, vol. 10, no. 9, pp. 487–495, Sep. 2013, doi: 10.1080/15459624.2013.818231.

[6] M. B. Miles and A. M. Huberman, "Drawing Valid Meaning from Qualitative Data: Toward a Shared Craft," *Educational Researcher*, vol. 13, no. 5, pp. 20–30, May 1984, doi: 10.3102/0013189X013005020.

[7] N. Carter, D. Bryant-Lukosius, A. DiCenso, J. Blythe, and A. J. Neville, "The Use of Triangulation in Qualitative Research," *Oncology Nursing Forum*, vol. 41, no. 5, pp. 545–547, Sep. 2014, doi: 10.1188/14.ONF.545-547.

[8] A. Susanto, M. Tejamaya, R. N. Wulan, and E. K. Putro, "Chemical Health Risk Assessment (CHRA) in a Wet Assay and Fire Assay Laboratory (WAFAL)," *Act Scie Medic*, vol. 4, no. 10, pp. 91–101, Sep. 2020, doi: 10.31080/ASMS.2020.04.0746.

[9] L. Hodson and M. Hull, "Building a Safety Program to Protect the Nanotechnology Workforce: A Guide for Small to Medium-Sized Enterprises," U.S. Department of Health and Human Services, Public Health Service, Centers for Disease Control and Prevention, National Institute for Occupational Safety and Health, Mar. 2016. doi: 10.26616/NIOSHPUB2016102.

[10] E. M. Osman, "Environmental and Health Safety Considerations of Nanotechnology: Nano Safety," *BJSTR*, vol. 19, no. 4, Jul. 2019, doi: 10.26717/BJSTR.2019.19.003346.

[11] V. Gharibi, A. Barkhordari, M. Jahangiri, M. Eyvazlou, and F. Dehghani, "Semi-Quantitative Risk Assessment of Occupational Exposure to Hazardous Chemicals in Health Center Laboratories (Case Study)," *Shiraz E-Med J*, vol. 20, no. 10, Sep. 2019, doi: 10.5812/semj.86764.

[12] M. I. Greenberg, D. C. Cone, and J. R. Roberts, "Material Safety Data Sheet: A Useful Resource for the Emergency Physician," *Annals of Emergency Medicine*, vol. 27, no. 3, pp. 347–352, Mar. 1996, doi: 10.1016/S0196-0644(96)70272-X.

[13] NIOSH, "General Safe Practices for Working with Engineered Nanomaterials in Research Laboratories," U.S. Department of Health and Human Services, Public Health Service, Centers for Disease Control and Prevention, National Institute for Occupational Safety and Health, May 2012. doi: 10.26616/NIOSHPUB2012147.

[14] M. G. M. Berges, R. J. Aitken, S. A. K. Read, K. Savolainen, M. Luotamo, and T. Brock, "Risk Assessment and Risk Management," in *Handbook of Nanosafety*, Elsevier, 2014, pp. 279–326. doi: 10.1016/B978-0-12-416604-2.00008-1.

[15] C. Geraci *et al.*, "Perspectives on the Design of Safer Nanomaterials and Manufacturing Processes," *J Nanopart Res*, vol. 17, no. 9, p. 366, Sep. 2015, doi: 10.1007/s11051-015-3152-9.

[16] A. Garcia, A. Eastlake, J. L. Topmiller, C. Sparks, K. Martinez, and C. L. Geraci, "Nano-Metal Oxides: Exposure and Engineering Control Assessment," *Journal of Occupational and Environmental Hygiene*, vol. 14, no. 9, pp. 727–737, Sep. 2017, doi: 10.1080/15459624.2017.1326699.

[17] SLAC National Accelerator Laboratory, "Nanomaterial Safety Plan," Environment, Safety & Health Division, 2017. [Online]. Available: www-group.slac.stanford.edu/esh/eshmanual/references/chemsafetyPlanNano.pdf

[18] NIOSH, "Protecting Workers During Intermediate and Downstream Processing of Nanomaterials," U.S. Department of Health and Human Services, Public Health Service, Centers for Disease Control and Prevention, National Institute for Occupational Safety and Health, Mar. 2018. doi: 10.26616/NIOSHPUB2018122.

[19] G. H. Amoabediny *et al.*, "Guidelines for Safe Handling, Use and Disposal of Nanoparticles," *J. Phys.: Conf. Ser.*, vol. 170, p. 012037, May 2009, doi: 10.1088/1742-6596/170/1/012037.

12 Fabrication and Characterization of Dye Sensitized Solar Cell in Various Metal Oxide Structure

Herlin Pujiarti,[1,2*] *Nabella Sholeha,*[1]
and Nadiya Ayu Astarini[1]
[1] Physics Department, Universitas Negeri Malang, Indonesia
[2] Centre of Advanced Materials for Renewable Energy
(CAMRY), State University of Malang, Indonesia
[*] Corresponding author: Herlin Pujiarti, herlin.pujiarti.
fmipa@um.ac.id

CONTENTS

DOI: 10.1201/9781003320746-12

12.1 INTRODUCTION

Renewable energy (RE) and sustainable supply chain management (SSCM) play substantial roles in the global energy industry. SSCM has been studied to build concepts related to the supply chain (Carter et al., 2019; Seuring & Müller, 2008). New renewable energy (NRE) has become an important issue in the green economy because the use of fossil fuels has decreased due to greenhouse gas (GHG) emissions and low costs (Al-Mulla et al., 2013). Power generation is another sector for which green technology can be generated. The distribution of green technologies, such as solar photovoltaic, biogas production, wind power, etc., has practically proven that these technologies can contribute the opportunities for people and can be applied to contribute to the energy solutions successfully. The category of RE that is important and widely used in the world is solar energy. It can be transformed into electricity in two ways: photovoltaic (PV devices or solar cells) and thermal power generation/solar electricity. Tropical countries get great advantages in various types of energy (Salunkhe & Shembekar, 2012).

Dye-sensitized solar cell (DSSC) is one of the solar cell devices that is being developed by researchers today because of its simple fabrication, environmental friendliness (Shakeel Ahmad et al., 2017), high efficiency, and the ability to convert solar energy into electricity (Diantoro, Suprayogi et al., 2019). O'Regan and Gratzel first reported with a DSSC efficiency of 7.1% in 1991, which increased in 2014 by 13% (Solehudin et al., 2020). In general, DSSC consists of four components, namely, a photoanode, a dye, an electrolyte, and a counter electrode in the form of a platinum-coated conductive glass (Andualem & Demiss, 2018). Currently, many studies focus on semiconductor materials as DSSC photoanodes such as TiO_2, Fe_2O_3, SnO_2, ZnO, and Nb_2O_5 (Diantoro, Maftuha et al., 2019; Dou et al., 2015). TiO_2 or titanium dioxide is one of the materials that is often used as a DSSC photoanode because it has a good bandgap, surface structure, particle size, porosity, and film thickness (Chamanzadeh et al., 2021). Therefore, the DSSC photoanode was developed using another semiconductor material, namely, zinc oxide (ZnO). ZnO is a semiconductor material that has a bandgap of 3.37 eV (Hezam et al., 2018) and a high electron mobility (200 $cm^2/V/s$) (Shah et al., 2021).

12.2 DYE-SENSITIZED SOLAR CELLS

DSSC is a thin-film photovoltaic technology. The photovoltaic performance of DSSC relies on light harvesting efficiency (LHE) and charge transfer. This parameter is affected by the type, morphology, composition, and method of taking the photoanode material. In this case, all the parameters and their impact on the performance of the DSSC will be discussed in detail.

12.2.1 DSSC Structure

The DSSC involves conductive substrates, metal oxide semiconductor, dye, electrolyte, and catalyst material. The basic structure and components of DSSC are shown in Figure 12.1 (Karim et al., 2019).

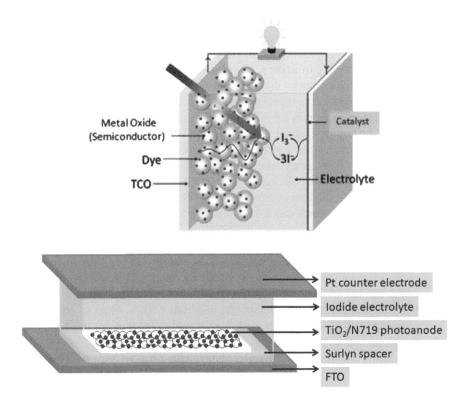

FIGURE 12.1   (a) Structure of DSSC (Yang et al., 2013). (b) Component of DSSC.

1. Conductive Substrate
   The substrate is a thin layer that is used as a current collector and a structure that supports the DSSC semiconductor layer.
2. The main material of the photoanode uses a metal oxide semiconductor material that functions as an electron collector in the DSSC. Morphology and bandgap affect the overall efficiency performance of the system. In this case, morphology and optical properties play a role in transferring electrons to external circuits and media for absorbing dyes. The characteristics that need to be considered to produce metal oxide semiconductor materials are:
   i. Photocorrosion resistance
   ii. High surface area to absorb dye
   iii. Electron acceptability, rough surface, high light diffusion
   iv. High bandgap
   v. Good adhesion to conductive substrates
      Popular metal oxide semiconductors that are often used for DSSC photoanodes are TiO_2, ZnO, and SnO_2. Characteristics that support good photon absorption are semiconductors having a bandgap higher than 3 eV. That way, the semiconductor can become a photoanode that functions well to absorb electrons properly.

3. Photo sensitizer

Photosensitizer is a crucial part of DSSC. The mechanism related to the role of the sensitizer is the adsorption of electrons from sunlight and the mechanism of electrons moving from the valence band to the conduction band of the semiconductor. Characteristics that need to be owned by dyes are:

  i. Absorb light in the spectrum in the ultraviolet (UV) region
 ii. Adhesion to support absorption in photoanode semiconductors
iii. The lowest unoccupied molecular orbital (LUMO) is larger than the conduction band in order to speed up the electron charge transport
 iv. Dye must be properly regenerated
  v. Has good thermal, electrochemical, and photostability properties
     Among the good photosensitizers for DSSC (N3, N719, and N749), dyes with Ru-complex showed a conversion efficiency exceeding 10%, thereby increasing the performance of DSSC.

4. Electrolyte

The performance of DSSC depends on the stability of the electrolyte which plays a role in regenerating the dye after the electron mechanism is excited to the conduction band of the semiconductor. Another role of the electrolyte is to transfer the holes to the counter electrode (Pt). In order to achieve the characteristics of electrolytes with high stability, electrolytes must have the following characteristics:

  i. Achieve high conductivity and voltage characterized by a positive redox potential
 ii. Low viscosity for fast electron transfer
iii. Interface contacts with semiconductors, counter electrodes, and nanocrystals.
 iv. Do not absorb sunlight in the visible light region

5. Counter Electrode

The counter electrode prepared for DSSC uses a platinum-based catalyst material. The catalyst layer serves to accelerate the reduction reaction when the electrolyte diffuses toward the counter electrode. The platinum (Pt) catalyst has a high vapor density, transparency, and good catalytic properties for supporting glass which is not a good electrode. However, Pt has several disadvantages, namely, high cost and poor photocorrosion. Therefore, another alternative is sought as a solution to the Pt deficiency, namely, by using a carbon allotrope-based material. However, the Pt photovoltaic counter electrode performance is good compared to other alternative materials.

12.2.2   MECHANISM

Interfacial electron transfer kinetics play an integral part in determining the photoelectric conversion efficiency (PCE) of Dye Sensitized Solar Cell (DSSC). DSSC components from left to right are platinized SnO_2 conductor, iodide, dye molecule, TiO_2 and SnO_2 conductor. Figure 12.2 presents a schematic of the charge transfer mechanism in DSSC.

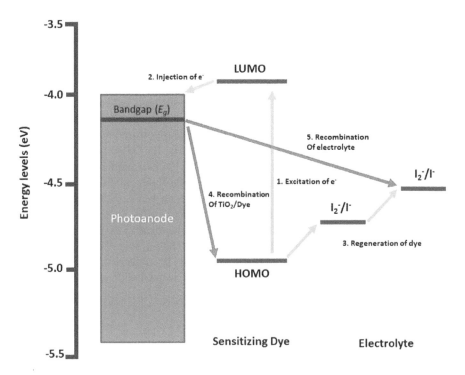

FIGURE 12.2   Transfer mechanism in DSSC (Holmberg et al., 2014).

The DSSC mechanism begins with the absorption of light by molecules in the dye. The dye molecules are integrated into the conduction band of the semiconductor. Excited electrons are injected into a conductive substrate. Next, the iodide ion is regenerated by the dye to oxidized triiodides ion. Platinum catalysts play a role in reduction reactions to speed up the passage of molecules. Several things need to be considered so as not to reduce the photovoltaic performance, namely, avoiding the diffusion of electrons from the photoanode back to the oxidized dye and avoiding electrons returning from the excited state to the ground state. In order for energy to continuously flow in the DSSC, it is necessary to pay attention to several things.

i. Photons must be scattered continuously.
ii. Shows no electrons trapped in the conduction band.
iii. Shows the even distribution of molecules in the semiconductor.
iv. Electrons in the conduction band must transition well with oxidized redox ions.
v. Electrons must be trapped with triiodide molecules from electrolyte.

12.3 TITANIUM DIOXIDE

The photoanode acts as an electron collector. Mesoporous semiconductor material for DSSC photoanode plays an important role in photovoltaic performance, which affects the electron collection process that experiences photoexcitation from dye molecules. Key parameters of harvesting depend on a lot of light on the surface of the photoanode film. The photoanode performance depends on the microstructure, crystallinity, morphology, porosity, surface area, and conduction band of the semiconductor (Karim et al., 2019). Therefore, photoanodes must have the main characteristics in order to produce optimal device performance, namely, high surface area to absorb a lot of dyes, high photocorrosion resistance, light scattering ability, light roughness, and good adhesion with conductive substrates. These characteristics support the performance of DSSC to get high efficiency conversion.

TiO_2 was chosen as the best semiconductor material for DSSC because it showed an efficiency of 14.5% (Kandasamy et al., 2022). TiO_2 is a semiconductor material against metal oxides with characteristics that are light stable, nontoxic, inexpensive, and widely available with energy (Eg) = 3 Ev (Karim et al., 2019). Based on the crystal form, TiO_2 is divided into three forms, namely, rutile, anatase, and brookite. The crystalline form of anatase TiO_2 has an optimal bandgap (3.20 eV), surface area, and a higher electron diffusion coefficient than rutile (Karim et al., 2019). Meanwhile, TiO_2 with brookite phase is difficult to synthesize and is not recommended for DSSC photoanode. As a result, TiO_2 anatase was suggested as the main constituent of DSSC photoanode.

12.3.1 MATERIALS AND METHODS

The TiO_2 photoanode layer consists of a blocking layer (BL-1 Blocking Layer, Greatcell), a mesoporous layer (18 NR-T Titania paste, Greatcell), and a reflective layer (Ti-Nanoxide R/SP, Solaronix). The blocking layer was deposited using the spin coating method (3,000 rpm, 30 seconds) and then heated gradually at temperatures of 100, 300, and 500°C for 15, 15, and 30 minutes. Meanwhile, the mesoporous layer and reflective layer were deposited using the screen-printing method with mesh T-61 and then heated gradually at temperatures of 100, 300, and 500°C for 15, 15, and 30 minutes. TiO_2/rGO was synthesized with TiO_2 paste and 0.5 wt% coated with one and two layers of photoanode to determine the best potential of rGO-based coating for DSSC photoanode. At the end of the TiO_2 film production, post-treatment was carried out by immersing in 2-propanol (Merck) and titanium (IV) (triethanolaminato) isoproxide solution (Sigma Aldrich) in an oven at 75°C. DSSC sandwich structure is composed of N719 as photosensitizer, Mesolyte as liquid electrolyte, and Pt as counter electrode. Then the TiO_2 film will be tested for its physical characteristics using XRD and SEM, while the performance of DSSC is characterized by using I-V with solar simulator and electro impedance spectroscopy (EIS). Figure 12.3 shows the fabrication of DSSC.

12.3.2 DISCUSSION

12.3.2.1 Microstructure

The photoanode performance depends on the purity of the metal oxide, thereby increasing the photoanode performance for efficient DSSC. The mechanical properties of nanomaterials are obtained from its crystallinity. An increase in mechanical strength can occur if there is a slight possibility of imperfections or crystal defects in

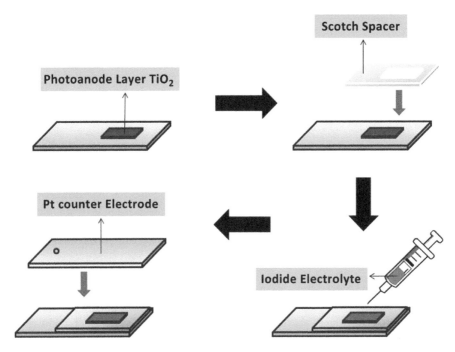

FIGURE 12.3 Fabrication of DSSC.

the material of the components of the system. According to thermodynamic theory, imperfections in crystals are highly possible. The small particle size reduces the possibility of such imperfections. The photoanode acts as a charge collector for electrons. The overall performance of the DSSC device depends on the purity of the metal oxide which contributes to the higher electron charge collection. The semiconductor TiO_2 has been investigated as a metal oxide with potential nontoxic, low-cost and widely available natural characteristics (Karim et al., 2019). TiO_2 is found in three different crystalline phases, namely, rutile, anatase, and brookite.

The X-ray diffraction (XRD) pattern of the TiO_2 film is shown in Figure 12.4(a). Based on the diffraction pattern, all phases were mainly in the TiO_2 plane anatase phase (COD 2310710), which was located at $2\theta = 25.3°$, $37.8°$, $48.1°$, $53.9°$, and $55.0°$. The characteristic peaks of the anatase phase correspond to (101), (004), (200), (211), and (204). Figure 12.4(b) shows the structure of TiO_2 with a tetragonal crystal system and a symmetry space group I 41/a m d illustrated using VESTA.

X-ray diffraction spectroscopy was used to investigate the crystalline phase, crystal size, and lattice parameters of the sample. Figure 12.4 shows the XRD pattern of the TiO_2 film synthesized using the screen-printing method. The diffraction peaks closely match the teragonal TiO_2 anatase phase. Sharp crystal peaks with high intensity indicate that TiO_2 produces a crystalline phase. The crystalline phase of TiO_2 anatase was extensively used as the main constituent for the photoanode characteristics of DSSC (Karim et al., 2019). TiO_2 anatase has a large surface area, optimal bandgap (3.20 eV), and a high electron diffusion coefficient. Based on XRD data, it can be concluded that

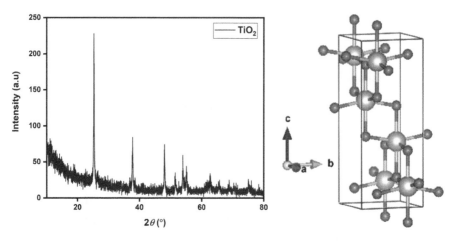

FIGURE 12.4   X-ray diffraction of TiO_2: (a) XRD and (b) crystal structure.

the sample formed pure phase anatase TiO_2 structure and no additional peaks were observed in the diffraction pattern. Figure 12.4 shows the XRD pattern of bare TiO_2.

XRD data were analyzed using Origin 2018 software with the Pseudovoigt1 function which corresponds to the shape of the XRD diffraction peak pattern. The average crystal size was calculated using the Scherer equation according to Equation 12.1.

$$D = \frac{0.89\lambda}{\beta_{hkl} \cos\theta} \qquad (12.1)$$

where D is the crystal size, k is the form factor, λ is the X-ray wavelength (1.5406), is the full half-width of the maximum (FWHM) diffraction peak and the diffraction angle. From Figure 12.5, we know that the FWHM of bare TiO_2 is 0.18766 and the crystallite size of bare TiO_2 is 48.2 nm. Figure 12.5 shows the measurement of FWHM via crystal plane diffraction (101).

The XRD pattern of the two layer of TiO_2/rGO film is shown in Figure 12.6. Based on the diffraction pattern, all phases were mainly in the TiO_2 plane anatase phase (COD 2310710) which was located at $2\theta = 25.3°$, $37.8°$, $48.1°$, $53.9°$, and $55.0°$. The characteristic peaks of the anatase phase correspond to (101), (004), (200), (211), and (204). From Figure 12.6(b), we know that the FWHM of TiO_2/rGO is 0.283 and the grain size of TiO_2/rGO is 2.71 μm.

12.3.2.2   Morphology

The morphology of the photoanode material was analyzed to review the characteristics of an efficient DSSC. TiO_2 particle size in the photoanode film affects the performance of solar cells. The large particle size distribution provides a small surface area for photosensitizer adsorption, thereby reducing the light absorbed by the photoanode in generating current from the DSSC cell. In contrast, photoanode films consisting

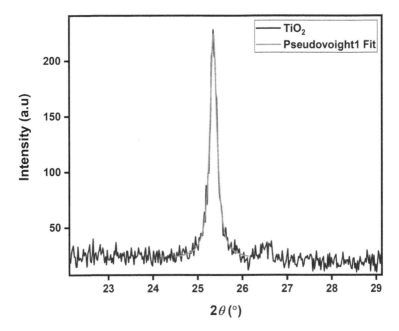

FIGURE 12.5 FWHM of TiO$_2$ film.

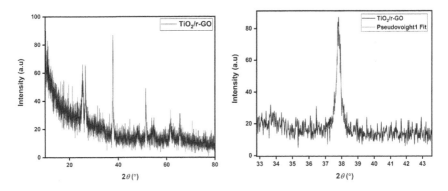

FIGURE 12.6 X-ray diffraction of TiO$_2$/rGO: (a) X-ray diffraction and (b) FWHM.

of small particles exhibit a larger number of grain boundaries, resulting in electron trapping when injected by excited photosensitizer molecules (Karim et al., 2019).

SEM images confirm the porous microstructure with a tetragonal TiO$_2$ matrix which confirms the XRD results in the previous characterization. The tetragonal microstructure of TiO$_2$ helps to increase the interface contact between the electrolyte and counter electrode. The homogeneous structure morphology offers a faster electron charging pathway and contributes significantly to the improvement of electron charge transport around the electrolyte/counter electrode interface. Porosity in the

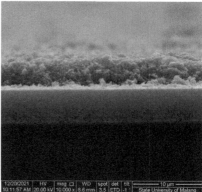

FIGURE 12.7   SEM image of TiO$_2$: (a) morphology and (b) cross section.

form of a tetragonal structure provides good stability conditions due to the presence of oxygen atoms at the edge of the surface which will provide surface sensitivity conditions for the DSSC photoanode (Amir-Al Zumahi et al., 2021). The third layer of the photoanode is the reflective layer. The light-scattering layer plays a major role in improving the light absorbing properties of the photoanode. The reflective layer is deposited on top of the mesoporous layer with a larger particle size. Figure 12.7 shows the SEM image of bare TiO$_2$.

Material properties can be analyzed through the size, shape, or area of agglomeration. The particle size determines the recency of the better properties. Particle sizes less than 100 nm in semiconductor materials, metals, and molecular crystals have lower melting temperatures than other solids. In general, a decrease in melting point indicates that the surface energy increases as the particle size decreases. In general, photoanodes with a high surface area reduce their ability to capture light due to their larger grain boundaries. But on the other hand, semiconductors with larger particle sizes serve to capture more incoming light and scatter more photons at the photoanode. Surface area results in greater surface energy. The large surface energy can improve the application carrier properties. The high surface energy is indicated by the small grain size.

The mesoporous photoanode consists of a metal oxide semiconductor material. The mesoporous semiconductor at the photoanode has an important role in the collection of excited electrons from the dye molecule. The mesoporous structure of TiO$_2$ has good dye adsorption ability (Karim et al., 2019). The key parameter for light harvesting depends on the amount of dye adsorption on the semiconductor film. The TiO$_2$ semiconductor is applied to most DSSC photoanodes due to its large specific surface area (Kandasamy et al., 2022). The higher the surface area of the material, the greater the absorption potential of the photosensitizer. Figure 12.8 shows the SEM of TiO$_2$/rGO.

12.3.2.3   Electrochemical Properties

The charge transfer mechanism of the DSSC interface was investigated by EIS measurements measured for the applied open circuit potential (OCP) under light

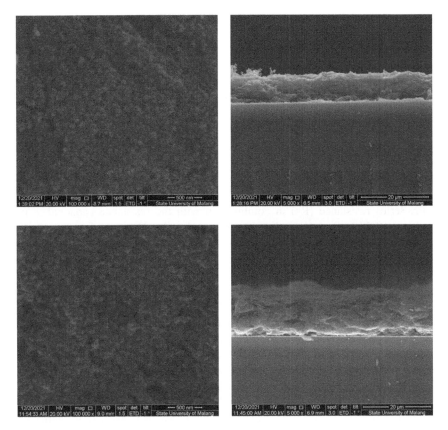

FIGURE 12.8   SEM of TiO_2/rGO: (a) morphology of 1 layer of TiO_2/rGO, (b) cross section of 1 layer of TiO_2/rGO, (c) morphology of 2 layer of TiO_2/rGO, and (d) cross section of 2 layer of TiO_2/rGO.

TABLE 12.1
EIS Performance Parameters on DSSC with Photoanode-Based Bare TiO_2

| | | | | | | CPE$_1$ | | CPE$_2$ | |
| | | | | | | Ym$_1$ (S ∘ s^a) | | Ym$_2$ (S ∘ s^a) | |
Sel	R$_s$ (Ω)	R$_1$ (Ω)	R$_2$ (Ω)	R$_3$ (Ω)	C$_1$ (mF)	× 10^{-3}	a$_1$	× 10^{-3}	a$_2$
Bare TiO_2	37.27	7.297	13.87	46.44	0.1419	1.156	0.5456	9.604	0.8730

conditions. Table 12.1 shows the EIS performance parameters on DSSC with different working areas. In general, the Nyquist plot on the DSSC shows three semicircles in the high-, medium-, and low-frequency regions. The first half circle in the high-frequency region represents the redox reactions in the electrolyte/Pt interface (Rct), the second half circle in the medium-frequency region corresponds to the charge transfer of the TiO$_2$/dye/electrolyte interface (Rrec), and the third half circle in the low-frequency region corresponds to Warburg diffusion resistance (Rw) of I-/I3 in electrolytes. The gradient occurring in the high-frequency region of the first semicircle contributes to the series resistance (Rs) due to the resistance in the fluorine thin oxide (FTO) conductive glass substrate (Subalakshmi & Senthilselvan, 2018). In addition, Cμ signifies the variation in the total chemical potential of the electrons in the TiO$_2$ photoanode. However, the Nyquist plot shows the behavior of the double semicircle due to the presence of electron transport resistance in the inner grains and particle boundaries on the surface of the TiO$_2$ film (Pujiarti et al., 2019).

The experimental data obtained are modeled with the equivalent circuit consisting of the series resistance (Rs) in combination with the resistance capacitance poly ethylene (RCPE). Based on the data shown by the Nyquist plot, two distinct semicircles were obtained, of which the mechanism is most clearly seen in the high-frequency region representing the charge transfer of the counter electrode (Naveen Kumar et al., 2020). In the low-frequency region, a semicircular curve is not fully visible, but this curve indicates a charge transport reaction that occurs at the TiO$_2$-dye interface with the electrolyte.

The electrochemical parameters and the change in charge transfer resistance at the TiO$_2$ DSSC photoanode are presented in Table 12.1. Based on the data shown in Table 12.1, the resistance in a cell with a working area of 0.5 × 0.5 cm^2 shows low resistance due to efficient charge collection. The increase in the resistance value is due to the presence of a number of grain boundaries that inhibit the electron jump and decrease the electrical conductivity. In general, the presence of charge-space polarization at low-frequency potentials tends to reduce the conductivity in the grain boundary region. However, the decrease in grain boundary resistance obtained in this study is caused by the wider active surface of the photoanode, which tends to increase the rate of charge transport and the possibility of recombination between paired electron holes rather than charge transport.

12.3.2.4 Electrical Properties

Photovoltaic current density (J-V) characteristics by anatase TiO$_2$-based DSSC were measured under a solar simulator lamp with an intensity of 100 mW/cm^2. In general, DSSC consists of several components arranged in a sandwich manner, namely, a conductive glass substrate, photoanode, photosensitizer, electrolyte, and counter electrode. The photovoltaic performance of DSSC depends on several factors, namely, electron collection at the photoanode, the efficiency of the charge absorption mechanism in photoanode light harvesting, electron scattering at the photoanode, and the efficiency of the electrolytic rapid reduction process at the counter electrode (Karim et al., 2019). Research with various sizes of photoanode working area was studied to determine the optimal condition of DSSC. The dye molecule is bonded to the metal oxide semiconductor via covalent bonds. The adsorbed dye absorbs solar radiation and the electrons in the dye become excited. Electrons tend to move from a lower energy level to a higher energy level. The highest occupied molecular orbital

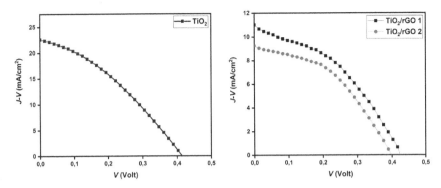

FIGURE 12.9 *J-V* characterization of DSSC: (a) photoanode-based bare TiO_2 and (b) photoanode-based TiO_2/rGO.

(HOMO) is considered a low-energy orbital, while the lowest vacant molecular orbital (LUMO) is considered a high-energy orbital. The electrons are then fed into the conduction band of the metal oxide semiconductor. They pass through a thin layer of mesoporous semiconductor and finally collect on a conductive substrate. The electrons from the photoanode then reach the capacitor electrolyte (CE) via an external circuit. Meanwhile, the electrolyte regenerates the oxidized dye and then diffuses toward the counter electrode. At the counter electrode, a reduction reaction occurs and the electrolyte is regenerated here. In this way, the DSSC work cycle is completed.

Figure 12.9 shows the J-V DSSC graph with different working areas. The electronic properties tested for DSSC photovoltaic parameters with different working areas are shown in Table 12.2. The conversion efficiency of DSSC with a photoanode-based bare TiO_2 is 4.67%, with Jsc = 22.6 ma/cm², Voc = 0.597 V, and FF = 0.346. The DSSC with the bare TiO_2 without other composite showed more optimum efficiency than the other comparison areas due to the increased Jc and Voc values. Otherwise, the conversion efficiency of DSSC with a photoanode based on TiO_2/rGO is 2.69%, with Jsc = 11.0 Ma/cm², Voc = 0.618 V, and FF = 0.396. The increase in conversion efficiency was better by 4.67% with an increase in Jsc (18.3 Ma/cm²) and Voc (0.591 Volt) than other cells due to higher dye adsorption, larger surface area, and minimum recombination. Increased efficiency is also associated with higher light scattering ability in cells with smaller working area (Naveen Kumar et al., 2020). The short-circuit current density (Jsc) depends on two factors, namely, the amount of dye loading and the electron injection rate at the dye/TiO_2 interface in photoanode based on TiO_2/rGO. Voc is determined by the gap between the energy level and the redox potential.

There are several factors that hinder the photovoltaic performance of DSSC, namely, the charge transport mechanism at the photoanode interface which is detrimental to the potential of DSSC. An important parameter is the conductivity of the substrate. The conductive transparent substrates used for DSSC are FTO and ITO. One of the signs of a decrease in DSSC performance is ineffective light scattering in

TABLE 12.2
Parameter Performances of DSSC

	J_{sc} (mA/cm²)	V_{oc} (V)	FF	Efficiency (%)
Bare TiO$_2$	22.6	0.597	0.346	4.67
TiO$_2$/rGO 1	11.0	0.618	0.396	2.69
TiO$_2$/rGO 2	9.24	0.572	0.431	2.28

the photoanode, that is, the optical properties are poor and the current density is low. TiO$_2$ photoanodes with poor light scattering are associated with light heat electron (LHE) and light scattering electron (LSE) (Lai et al., 2019), which can be overcome by changing the morphology of TiO$_2$ (nanowires, nanotubes, core shell structure, etc.) or TiO$_2$/carbon nanocomposites. DSSC devices with liquid electrolyte show the factors that influence the reduced stability due to system leakage. Therefore, solid-state electrolyte is an alternative to improve the stability of DSSC. The DSSC component that is experiencing challenges in finding alternatives is the counter electrode catalyst. Because the established Pt counter electrode is expensive and photocorrosion is not good, research and development is needed to produce a cheap and efficient catalyst. In the period of research work, conductive polymer nanocomposites can be an alternative as a catalyst to replace Pt.

12.4   ZINC OXIDE

Zinc Oxide is a semiconductor material that has a bandgap of 3.37 eV (Hezam et al., 2018) and a higher electron mobility (200 cm²/V/s) (Shah et al., 2021). Various ZnO morphologies were applied to DSSC, such as thin films, nanotubes, nanoflowers, and nanorods, and the most widely used is the nanorod (Shah et al., 2021). This is because ZnO with nanorod morphology has low electron hopping (Dou et al., 2015), high electron mobility, good chemical and thermal stability, high specific surface area, abundant availability in nature, and is environmentally friendly (Diantoro, Maftuha, et al., 2019). However, the photoconversion efficiency of DSSC with ZnO nanorod-based photoanode is usually lower than that of DSSC with TiO$_2$-based photoanode (Kumar et al., 2021). Therefore, various efforts have been made to increase the efficiency of DSSC based on ZnO nanorod, including by varying the thickness of the seed layer with the spray method and the growth time of ZnO nanorod, and the concentration of its precursor (Abdulrahman et al., 2020). This affects the structure, morphology, and optical properties of ZnO nanorods (Abdulrahman et al., 2020). Thambidurai et al. succeeded in synthesizing ZnO nanorods by varying the concentration of precursors for DSSC applications and obtained diameters and average lengths ranging from 100 to 135 nm and 1 to 3 m, while the absorbance was shown in the range of 470–510 nm (Thambidurai et al., 2013). This shows that high absorbance results in high efficiency due to higher surface area, and high efficiency is shown by DSSC with a molar concentration of 0.07 M of 0.71% (Thambidurai et al., 2013).

12.4.1 Materials and Methods

The ZnO nanorods on FTO was grown by hydrothermal synthesis method with different concentrations of ZnO nanoseed. The first step is washing the FTO substrate using foam solution, aquades, and acetone. The second step is preparing a ZnO nanoseed 0.05 M, 0.075 M, and 0.1 M of zinc acetate dihydrate (Pro Analyst, Merck) in ethanol (Pro Analyst, Merck). Then, the ZnO nanoseed solution was deposited on the FTO substrate by spin coating method at a speed of 2,000 rpm for 20 seconds, and then heated gradually, 100°C (15 minutes), 300°C (15 minutes), and 500°C (30 minutes). The third step is preparing the growth solution, zinc nitrate tetrahydrate (Pro Analyst, Merck), and hexamethylenetetramine (Pro Analyst, Merck) with the molarity ratio of 1:5 dissolved in aquades. In the fourth step, the substrate that has been deposited in the second step is immersed in a solution for the growth of ZnO nanorods; this method is known as hydrothermal method with 6 hours process. Finally, the samples are heated gradually as before.

Furthermore, for assembly into DSSC, the finished ZnO nanorod photoanode was first immersed in 0.1 M quercetin dye for 24 hours in the dark. After being given a surlyn-60 spacer and assembled with a counter electrode, then, Pt. injected with mesolyte electrolyte.

The XRD pattern was obtained by PANalytical X'Pert PRO using a Cu-Kα radiation source and the diffraction angle in the range of 10–90°. Morphological images and compound content were obtained by SEM-EDX FEI INSPECT-S50 type. The optical properties were carried out by UV-Vis UV-2600 type. The I-V characteristic was carried out by Keithley 6517B under the solar simulator (LS 150, Azet Technologies, Australia) with the intensity of 100 mW/cm^2.

12.4.2 Discussion

12.4.2.1 Microstructure

Figure 12.10 shows the XRD characterization of the ZnO nanorods thin film samples grown on the FTO substrate. ZnO with a wurtzite phase was formed as shown by a high peak at (002) at an angle of 2θ at 34°. This experimental phase was matched through the COD-9004179 in the Crystallography Open Database. This could be due to the duration of nanorod formation through the hydrothermal method which lasted for 6 hours. Some studies have stated that the increasing of duration in crystal growth will affect the increasing of crystallinity level. For epitaxial ZnO at different hydrothermal durations, it increases with the intensity of hydrothermal duration. This is indicated by the crystallinity of 52.38%.

12.4.2.2 Morphology

Furthermore, the morphological structure with a molarity variation of 0.05 M, 0.075 M, and 0.01 M in the ZnO nanoseed solution is shown in Figure 12.11. From the SEM characterization, it can be seen that all of the mappings of the nanorod tip are facing upward with a hexagonal plane shape. There are many shapes that clearly show it is hexagonal, but some are agglomerated or stacked between one rod and another. The

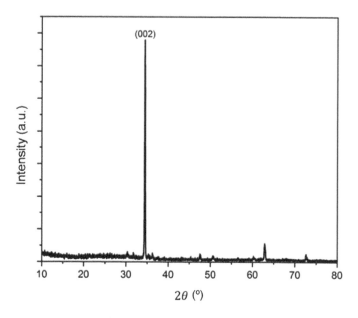

FIGURE 12.10   Diffractogram of ZnO nanorods-FTO substrate.

FIGURE 12.11   Morphology of ZnO nanorods with various molarity of Zn nanoseed-FTO substrate: (a) 0.05 M, (b) 0.075 M, and (c) 0.1 M.

morphological structure carried out by mapping will be reviewed again regarding the particle size and porosity of the thin film.

The morphological structure chosen to analyze this section is by selecting an image that looks clearly hexagonal, randomly. From an analysis and measurement of SEM images using imageJ, the highest average of the variations carried out is ZNR-0.05. In this condition, the particle size obtained is 841 nm with the smallest porosity of all, which is 58.3%. Meanwhile, for the thickness of the thin film, all of them have almost the same value, which is around 1.53 μm. On the other hand, these two, the average particle diameter and porosity, are also interconnected; the higher the average particle diameter, the higher the porosity value, as shown in Table 12.3.

TABLE 12.3
Diameter of ZnO Nanorods-FTO Substrate

Parameter	ZNR-0.05	ZNR-0.075	ZNR-0.1
Mean diameters (nm)	841	593	461
Porosity (%)	58.3	60.4	60.8

TABLE 12.4
Bandgap Energy of ZnO Nanorods-FTO Substrate

Sample	Band Gap Energy (eV)
ZNR-0.05	3.79
ZNR-0.075	3.21
ZNR-0.1	3.16
ZNR-0.05/dye	3.20
ZNR-0.075/dye	3.13
ZNR-0.1/dye	2.63

12.4.2.3 Optical Properties

The result of UV-Vis characterization is calculated the direct bandgap energy. The samples tested were samples with dye and without dye. The dye used is natural dye quercetin. Natural dye quercetin is yellow and dye loading is carried out for 24 hours at room temperature. From each sample tested, it was found that the bandgap energy decreased when the molarity was added to the Zn nanoseed solution; as shown in Table 12.4, the smallest bandgap energy was at ZNR-0.1, which was 3.16 eV, which decreased when quercetin dye was loaded and resulted in a bandgap energy of 2.63 eV. This decrease in bandgap energy in this sample has the potential for photovoltaic device applications, such as DSSC, which will be discussed later.

12.4.2.4 Electrical Properties

The J-V characteristics of the ZnO nanorods sample with variations of 0.05 M, 0.075 M, and 0.1 M nanoseed are shown in Figure 12.12.

The sample was prepared by combining DSSC parts, namely, photoanode, 0.1 M quercetin dye, electrolyte (Mesolyte), and counter electrode (Pt). The tested sample showed an increase in conversion efficiency with every increase in the molarity of the Zn nanoseed concentration. The bandgap is the minimum amount of energy required for electrons to escape from their bound state. The narrower the distance, the faster the electrons move, thereby improving their energy conversion performance. The photovoltaic parameters of the DSSC are presented in Table 12.5 in detail.

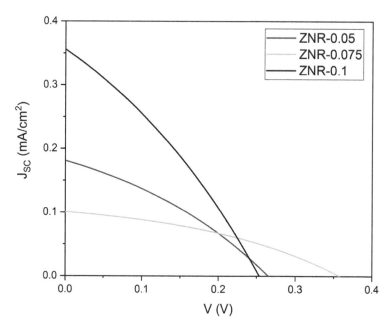

FIGURE 12.12   *J-V* characteristics for various molarities of nanoseed.

TABLE 12.5
Photovoltaic Parameter of DSSC

Sample	I_{sc} (mA)	J_{sc} (mA/cm²)	I_{max} (mA)	V_{max} (V)	V_{oc} (V)	FF
ZNR-0.05	0.09	0.11	0.04	0.20	0.30	0.30
ZNR-0.075	0.09	0.09	0.05	0.26	0.36	0.39
ZNR-0.1	0.15	0.22	0.11	0.16	0.29	0.38

12.5   CONCLUSION

Samples prepared with various molarities in the preparation of ZnO nanoseeds showed an increase in porosity in their morphological structure which then affected their optical and electrical properties. With the energy value of the bandgap getting narrower for each increase in molarity, the electrical properties showed an increase in the conversion of DSSC performance using quercetin dye, resulting in an efficiency of 0.025%. The best performance of DSSC is photoanode based on TiO_2 (4.67%). Photoanode TiO_2/rGO shows good characteristics for DSSC-based carbon for green energy (2.69%).

ACKNOWLEDGMENT

The authors would like to thank PNBP UM for the funding support through research grant number 5.3.484/UN32.14.1/LT/2021.

REFERENCES

Abdulrahman, A. F., Ahmed, S. M., Ahmed, N. M., & Almessiere, M. A. (2020). Enhancement of ZnO nanorods properties using modified chemical bath deposition method: effect of precursor concentration. *Crystals*, *10*(5), 386. https://doi.org/10.3390/cryst10050386

Al-Mulla, A., Maheshwari, G. P., Al-Nakib, D., ElSherbini, A., Alghimlas, F., Al-Taqi, H., & Al-Hadban, Y. (2013). Enhancement of building operations: A successful approach towards national electrical demand management. *Energy Conversion and Management*, *76*, 781–793. https://doi.org/10.1016/j.enconman.2013.07.080

Amir-Al Zumahi, S. M., Arobi, N., Mahbubur Rahman, M., Kamal Hossain, M., Ara Jahan Rozy, M., Bashar, M., Amri, A., Kabir, H., Abul Hossain, M., & Ahmed, F. (2021). Understanding the optical behaviours and the power conversion efficiency of novel organic dye and nanostructured TiO_2 based integrated DSSCs. *Solar Energy*, *225*, 129–147. https://doi.org/10.1016/j.solener.2021.07.024

Andualem, A., & Demiss, S. (2018). Review on dye-sensitized solar cells (DSSCs). *Journal of Heterocyclics*, *1*(1), 29–34. https://doi.org/10.33805/2639-6734.103

Carter, C. R., Hatton, M. R., Wu, C., & Chen, X. (2019). Sustainable supply chain management: continuing evolution and future directions. *International Journal of Physical Distribution & Logistics Management*, *50*(1), 122–146. https://doi.org/10.1108/IJPDLM-02-2019-0056

Chamanzadeh, Z., Ansari, V., & Zahedifar, M. (2021). Investigation on the properties of La-doped and Dy-doped ZnO nanorods and their enhanced photovoltaic performance of dye-sensitized solar cells. *Optical Materials*, *112*, 110735. https://doi.org/10.1016/j.optmat.2020.110735

Diantoro, M., Maftuha, D., Suprayogi, T., Reynaldi Iqbal, M., Solehudin, Mufti, N., Taufiq, A., Hidayat, A., Suryana, R., & Hidayat, R. (2019). Performance of *Pterocarpus Indicus Willd* leaf extract as natural dye TiO_2-Dye/ITO DSSC. *Materials Today: Proceedings*, *17*, 1268–1276. https://doi.org/10.1016/j.matpr.2019.06.015

Diantoro, M., Suprayogi, T., Taufiq, A., Fuad, A., & Mufti, N. (2019). The effect of PANI fraction on photo anode based on TiO_2-PANI /ITO DSSC with β-carotene as dye sensitizer on its structure, absorbance, and efficiency. *Materials Today: Proceedings*, *17*, 1197–1209. https://doi.org/10.1016/j.matpr.2019.05.345

Dou, Y., Wu, F., Mao, C., Fang, L., Guo, S., & Zhou, M. (2015). Enhanced photovoltaic performance of ZnO nanorod-based dye-sensitized solar cells by using Ga doped ZnO seed layer. *Journal of Alloys and Compounds*, *633*, 408–414. https://doi.org/10.1016/j.jallcom.2015.02.039

Hezam, A., Namratha, K., Drmosh, Q. A., Chandrashekar, B. N., Jayaprakash, G. K., Cheng, C., Srikanta Swamy, S., & Byrappa, K. (2018). Electronically semitransparent ZnO nanorods with superior electron transport ability for DSSCs and solar photocatalysis. *Ceramics International*, *44*(6), 7202–7208. https://doi.org/10.1016/j.ceramint.2018.01.167

Holmberg, S., Perebikovsky, A., Kulinsky, L., & Madou, M. (2014). 3-D micro and nano technologies for improvements in electrochemical power devices. *Micromachines*, *5*(2), 171–203. https://doi.org/10.3390/mi5020171

Kandasamy, M., Selvaraj, M., Alam, M. M., Maruthamuthu, P., & Murugesan, S. (2022). Nano-silver incorporated amine functionalized graphene oxide titania nanotube composite: a promising DSSC photoanode. *Journal of the Taiwan Institute of Chemical Engineers*, *131*, 104205. https://doi.org/10.1016/j.jtice.2022.104205

Karim, N. A., Mehmood, U., Zahid, H. F., & Asif, T. (2019). Nanostructured photoanode and counter electrode materials for efficient dye-sensitized solar cells (DSSCs). *Solar Energy*, *185*, 165–188. https://doi.org/10.1016/j.solener.2019.04.057

Kumar, V., Gupta, R., & Bansal, A. (2021). Hydrothermal growth of ZnO nanorods for use in dye-sensitized solar cells. *ACS Applied Nano Materials*, *4*(6), 6212–6222. https://doi.org/10.1021/acsanm.1c01012

Lai, C. W., Low, F. W., Siddick, S. Z. B. M., & Juan, J. C. (2019). Graphene/TiO$_2$ nanocomposites: synthesis routes, characterization, and solar cell applications. In E. Celasco, A. N. Chaika, T. Stauber, M. Zhang, C. Ozkan, C. Ozkan, U. Ozkan, B. Palys, & S. W. Harun (Eds.), *Handbook of Graphene* (1st ed., pp. 353–394). Wiley. https://doi.org/10.1002/9781119468455.ch64

Naveen Kumar, T. R., Yuvaraj, S., Kavitha, P., Sudhakar, V., Krishnamoorthy, K., & Neppolian, B. (2020). Aromatic amine passivated TiO$_2$ for dye-sensitized solar cells (DSSC) with ~9.8% efficiency. *Solar Energy*, *201*, 965–971. https://doi.org/10.1016/j.solener.2020.03.077

Pujiarti, H., Hidayat, R., & Wulandari, P. (2019). Enhanced efficiency in dye-sensitized solar cell by localized surface plasmon resonance effect of gold nanoparticles. *Journal of Nonlinear Optical Physics & Materials*, *28*(04), 1950040. https://doi.org/10.1142/S0218863519500401

Salunkhe, P. B., & Shembekar, P. S. (2012). A review on effect of phase change material encapsulation on the thermal performance of a system. *Renewable and Sustainable Energy Reviews*, *16*(8), 5603–5616. https://doi.org/10.1016/j.rser.2012.05.037

Seuring, S., & Müller, M. (2008). From a literature review to a conceptual framework for sustainable supply chain management. *Journal of Cleaner Production*, *16*(15), 1699–1710. https://doi.org/10.1016/j.jclepro.2008.04.020

Shah, S. A. A., Guo, Z., Sayyad, M. H., & Sun, J. (2021). Optimizing zinc oxide nanorods based DSSC employing different growth conditions and SnO coating. *Journal of Materials Science: Materials in Electronics*, *32*(2), 2366–2372. https://doi.org/10.1007/s10854-020-05001-2

Shakeel Ahmad, M., Pandey, A. K., & Abd Rahim, N. (2017). Advancements in the development of TiO$_2$ photoanodes and its fabrication methods for dye sensitized solar cell (DSSC) applications. A review. *Renewable and Sustainable Energy Reviews*, *77*, 89–108. https://doi.org/10.1016/j.rser.2017.03.129

Solehudin, S., Diantoro, M., & Hidayat, A. (2020). Analysis of nanoporous semiconductor film thickness effect on short circuit current density of β-carotene dye based DSSC: theoretical and experimental approach. *Journal of Physical Science and Engineering*, *4*(1), 1–7. https://doi.org/10.17977/um024v4i12019p001

Subalakshmi, K., & Senthilselvan, J. (2018). Effect of fluorine-doped TiO$_2$ photoanode on electron transport, recombination dynamics and improved DSSC efficiency. *Solar Energy*, *171*, 914–928. https://doi.org/10.1016/j.solener.2018.06.077

Thambidurai, M., Muthukumarasamy, N., Velauthapillai, D., & Lee, C. (2013). Chemical bath deposition of ZnO nanorods for dye sensitized solar cell applications. *Journal of Materials Science: Materials in Electronics*, *24*(6), 1921–1926. https://doi.org/10.1007/s10854-012-1035-8

Yang, X., Yanagida, M., & Han, L. (2013). Reliable evaluation of dye-sensitized solar cells. *Energy Environ. Sci.*, *6*(1), 54–66. doi: 10.1039/C2EE22998F

13 Characterizations of Amino-Functionalized Metal-Organic Framework Loaded with Imidazole

Mohd Faridzuan Majid,[1] Hayyiratul Fatimah Mohd Zaid,[2,3]* and Danialnaeem Emirzan Bin Mardani[3]
[1] Fundamental and Applied Sciences Department, Universiti Teknologi PETRONAS, Malaysia
[1,2] Centre of Innovative Nanostructures and Nanodevices (COINN), Universiti Teknologi PETRONAS, Malaysia
[2,3] Chemical Engineering Department, Universiti Teknologi PETRONAS, Malaysia
* Corresponding author: Hayyiratul Fatimah Mohd Zaid, hayyiratul.mzaid@utp.edu.my

CONTENTS

DOI: 10.1201/9781003320746-13

13.1   INTRODUCTION

13.1.1   THE WORLD OF NANOMATERIALS

Nanomaterials are substances that have a critical dimension between 1 and 100 nm. This type of material usually works outside of the molecular and bulk solid state due to its critical dimension in this range. In some textbooks or research articles, nanomaterials are sometimes referred to as *nanoscience, nanoparticle,* or *nanotechnology.* Nanomaterials can be classified based on their dimensionality— zero-dimensional (0D), one-dimensional (1D), two-dimensional (2D) and three-dimensional (3D)—which have different characteristics. Each of this has their advantages and specific applications. The low-dimensional nanomaterials would have a bigger total surface area, which may increase their chemical reactivity. One popular example of 0D nanomaterial is quantum dots, which is a synthetic nanoscale crystal that could transport electrons. 1D nanomaterial has one dimension which is outside the nanoscale system. This type of nanomaterials includes nanotubes, nanowire and nanorods. Similarly, a 2D nanomaterial has two dimensions which are out of the nanoscale range. This kind of nanomaterial is usually present in plate-like shapes, such as nanofilms, nanolayers, nanocoatings and graphene. Meanwhile, a 3D nanomaterial is not confined to the nanoscale dimension. The materials are hierarchical in structure and constructed from the smaller dimension of building blocks. The surface of this materials sometimes has a nature-inspired shape, such as flower petals, tree trunks and branches (Al-Akraa et al., 2017; Poerwoprajitno et al., 2020; P. Singh et al., 2016; Sreedharan et al., 2019). Because of the small size of nanomaterials, it offers a handful of benefits, especially the tunable design and manipulation of nanostructures which could accommodate various applications. Moreover, it allows the bonding of active raw materials which opens the possibility of synthesizing the desired materials. For example, in 3D nanomaterials such as zeolites, mesoporous silica and metal-organic framework (MOF), the presence of macropores and micropores in the framework is advantageous for carbon dioxide capture from the gas stream, which subsequently reduces the problems of greenhouse gas effect (Megías-Sayago et al., 2019, p. 2; Sadiq et al., 2020; Sneddon et al., 2014; Zagho et al., 2021). Additionally, the utilization of nanomaterials in energy storage system could improve the efficiency and economical value of the current methods of energy generation. Incorporating nanomaterials will reduce the diffusion length and increase the kinetics of energy storage system, which will eventually improve the charging-discharging process (Nagpure & Bhushan, 2016). Using these nanomaterials could also slow down the degradation of battery components which is caused by the contraction and expansion of volume in the electron transport process. This can enhance the working capacity, performance and longevity of energy storage device (Jaramillo-Cabanzo et al., 2021; Kim et al., 2021; M. Wang et al., 2021; Yaroslavtsev et al., 2015). However, there are some disadvantages that need to be considered when applying nanomaterials. Firstly, the exposure of nanomaterials can give a detrimental effect toward human health as some of the nanomaterials are toxic. Since nanomaterials are generally small, they can be easily entered into human body when subjected to long exposure. Experts suggested that any nanomaterials which is less than 10 nm behave similarly to gas and therefore can enter cell tissues and disturb the biological system (Dugershaw et al., 2020; Ganguly et al., 2018; van der Merwe & Pickrell, 2018; Vishwakarma et al., 2010). Secondly, the process of upscaling of the production of nanomaterials is still challenging due to the complex preparation of nanomaterial. Most of the studies are focusing on the lab scale synthesis of nanomaterials, which only yields a small amount of pure product. Moreover, there is no standardized procedure and specific guidance for the chemical process development due to the various ways of producing nanomaterials at a small scale. Furthermore, the stability of the raw materials and the final products of nanomaterials are still a difficult problem when placed in harsh conditions in a chemical reactor (Lin et al., 2019; Rhodes et al., 2019). Thirdly, the assessment of the green aspect of nanomaterials is still under debate. Because of the large manufacturing of the nanomaterials, issues such as the selection of raw materials, life cycle assessment, recyclability and reusability of nanomaterials need to be tackled. Some nanomaterials are hard to degrade, such as carbon nanotubes, magnetic nanoparticles and silica nanoparticles and will lead to waste management issues. Although there exist biodegradable nanomaterials, the stability and sensitivity of these materials especially in medicinal applications, such as drug delivery system, can be problematic for large-scale production and sterilization (Mitchell et al., 2021; Remya et al., 2022; Su & Kang, 2020). Despite all the challenges faced by nanotechnology industry, there is still room for improvement for these nanomaterials miniature as most of the powerful characterization tools are available to investigate and evaluate the full usage of nanomaterials (Akhtar & Ali, 2020; Hallan et al., 2021; Irshad et al., 2021; Kumar & Kumbhat, 2016; Salame et al., 2018). The design and analysis of green nanomaterials should take priorities for newly discovered materials and this leads us to the next subtopic which is the role of green chemistry in the quest of sustainable development.

13.1.2 Green Chemistry—Perspectives toward Sustainability

According to the United States Environmental Protection Agency (EPA), green chemistry is defined as the design of chemical substances and routes that decrease or exclude the usage or generation of harmful substances to human health and the environment. In other worlds, green chemistry is any process that governs the sustainability of the material from its initial production until it is no longer in use without disturbing the environment and its ecosystem. It is a systematic approach and should be the main guidelines for all chemical industry practitioners for material manufacturing. There are 12 principles of green chemistry: waste prevention, atom economy, usage of less hazardous chemical synthesis, production of safer chemicals, utilization of safer solvents and auxiliaries, energy-efficient process, usage of renewable raw materials, reduction of by-products and derivates, enhancement via catalysis, design for degradation, instantaneous pollution prevention and safer chemistry to prevent unwanted incidents (Anastas & Eghbali, 2010; Chen et al., 2020; Omran & Baek, 2022; Rogers & Jensen, 2019; Zimmerman et al., 2020). All these green chemistry principles when seriously implemented in chemical manufacturing will provide positive impacts towards the environment by reducing the environmental pollution. From the standpoint of economic development, a reduction in each manufacturing expenditure can be achieved by the utilization of small amount of solvents, simple synthesis approach and energy-efficient procedure. Minimum or zero waste from the production plant is another advantage and this will save a lot of time and energy to dispose the by-products. The practice of green chemistry in nanotechnology is now becoming a trend to address the issues faced by nanomaterials, especially in green synthesis of nanomaterials and nanotoxicology. Although exploitation of nanomaterials in green process such as in catalysis has already emerged as an eco-friendly approach for many organic and material synthesis, the green design of nanomaterial itself needs to be paid attention before we can fully utilized and integrate nanomaterials in our daily life. This is to ensure that the potential hazardous effects on human health can be reduced and to promote the replacement of nano-based product with a more environmentally benign material. The current green nanotechnology is focusing on environment remediation (water treatment, water cleaning technology and water disinfection, plastic removals from oceans and air pollution control), energy storage system and nanosensors. Among these applications, energy storage system is currently being one of the practical solutions to mitigate climate change. In one case study, it was found that penetration of energy storage in California and Texas could reduce 57% emissions of carbon dioxide, which exceeded the 55% carbon dioxide emissions from the 2030 Climate Target Plan sets by Europe (Arbabzadeh et al., 2019; Scarlat et al., 2015; 2018). Various energy storage systems are currently available in the market to power our daily lives, which include fossil fuel storage, mechanical-based, electrical and electromagnetic-based, biological, electrochemical, thermal and chemical storage systems. Among these, electrochemical energy storage system is widely used to generate electricity and storing energy for later use. Rechargeable battery is one of the popular battery systems that shape our daily task, allowing us to commute to work, making a phone call and having an online video conference. The three main types of rechargeable batteries are nickel-cadmium (NiCd), nickel-metal hydride (NiMH) and lithium-ion (Li-ion) batteries. The Li-ion technology is currently the major player in electrical energy storage as it has higher energy density than its predecessor. Incorporation of nanomaterials in the manufacturing of Li-ion components provides a significant enhancement in terms of its power and life cycle. It is also claimed to be less toxic as it uses non-toxic metals as electrodes. A Singapore-based company—the Green Li-ion—is the first battery manufacturer that produces green Li-ion battery that is recyclable without sacrificing the overall performance of the battery (Li et al., 2021; Thompson et al., 2020). This is one of the examples of material manufacturing which integrates the pillars of green chemistry.

13.2   LITHIUM-ION BATTERIES

Lithium-ion batteries (LIBs) have attracted great attention owing to their high energy density compared to other available rechargeable battery technologies. LIBs are commonly used as a power supply in portable electronic devices such as smartphone, laptop and tablet. Most recently, the popularity of LIBs has grown in electrical vehicles, military devices and aerospace industry. One of the main components of LIBs that plays an important role is electrolyte. An electrolyte will facilitate the movement of Li-ions from anode to cathode (discharging) and vice versa (charging) through a separator. When ion flows, free electrons are formed in the anode which creates charge at the positive current collector. This will generate electrical current flowing from the current collector through a plugged device and the negative current collector. Commercially available LIBs use organic solvent such as ethylene carbonate and dimethyl carbonate mixed with lithium hexafluorophosphate LiPF6 as LIBs electrolyte; however, they tend to explode when exposed to high temperature (Chatterjee et al., 2020; Roth & Orendorff, 2012). Although LIB technology provides great usefulness and portability to consumers, it carries the risk of explosions which may result in unwanted incidents.

Efforts have been made to address flammability concern in conventional LIBs including replacement of solution electrolyte with solid-state electrolyte (SSE). SSE has been introduced as an alternative ionic conductor due to its increased safety, greater thermal stability, nonvolatility and affordable power density and cyclability (Baumann et al., 2019). SSE possesses high lithium transference number compared to liquid electrolyte, which could circumvent ohmic polarization in LIBs. Example of SSE are inorganic solid electrolyte (ISE), solid polymer electrolyte (SPE) and composite polymer electrolyte (CPE) (Aziz et al., 2018; Liang et al., 2018; Tambelli et al., 2002; R. Zhao et al., 2020). They can provide fast lithium transportation but still suffer from mechanical strength and thermal stability. Searching for new classes of material is necessary to circumvent these issues.

13.3   METAL-ORGANIC FRAMEWORKS

MOF is a type of organometallic polymer structure which is composed of metal clusters linked with organic ligands. MOF is classified as a three-dimensional coordination compound because it contains a repeating complex that extends in x, y and z axes. It can be prepared via self-assembly which involves the crystallization of metal salt and organic ligand. The possible intermolecular forces in MOF are van der Waals forces, $\pi-\pi$ interactions, hydrogen bonding and stabilization of π-bond by polarized bonds from the synergistic interaction of metal and ligands. The geometry of MOF is adjustable depending on the type of metal nodes and length and functional groups of organic ligands. This porous crystalline material has a large total surface area (up to 6000 m^2/g), large pore volume (up to 90% free volume) and storage capacity (Fujie et al., 2014). Few studies have suggested that the functionality of MOF can be improved by loading guest molecules and modifying the organic linker (Fujie et al., 2014; Liu et al., 2015; A. Singh et al., 2017; Yoshida & Kitagawa, 2018). The insertion of these guest molecules could alter the phase behavior of the composite, eventually increasing its performance as adsorbent in gas purification and electrolyte for rechargeable battery. Shucheng, Zifeng and Yi reported the success of imidazole incorporation in UiO-67 as a potential superionic conductor candidate (Liu et al., 2015). No conductivity was observed for bulk imidazole; however, when incorporating imidazole into MOF, the dynamic motion was accelerated and hence it enhanced ionic conductivity. Xu et al. have reported the performance of cyanide-based MOF hybrid electrolytes (Xu et al., 2019). Two different cyanide ionic liquids, namely EMIM-SCN and EMIM-DCA, were introduced into MIL-101 to form two different electrolytes. The ionic conductivity recorded for SCN-type composite was 1.15×10^{-3}

S cm^{-1} at room temperature and could reach up to 6.21×10^{-2} at 150°C. Meanwhile, for DCA-type composite, the ionic conductivity was 4.14×10^{-4} at room temperature and increased to 2.45×10^{-3} at 150°C. Wang et al. highlighted the excellent proton conductivity when introducing amino acids onto UiO-66-NH$_2$ (S. Wang et al., 2021). Due to the uniform dispersion and interfacial effect of the nanocomposite MOF, more proton transfer sites can be provided, which improves the performance of proton conductivity (Moi et al., 2020). In this study, MIL-88 and UiO-66 were selected as MOF and both were functionalized with amino group at the organic linker side and loaded with imidazole as guest molecule. The morphology and molecular fingerprints were identified to ensure its potential application in LIB.

13.4 METHODOLOGY

13.4.1 CHEMICALS

Zirconium (IV) chloride, ZrCl$_4$, iron (III) chloride, FeCl$_3$, 4, 4'-biphenyl dicarboxylic acid, H2BPDC, 2-aminoterephthalic acid, C8H7NO4 and imidazole, C$_3$H$_4$N$_2$ were supplied by Sigma Aldrich. N, N-dimethylformamide, DMF and ethanol, CH$_3$CH$_2$OH were purchased from Avantis Laboratory. All chemicals were used without further purification steps.

13.4.2 SYNTHESIS OF IL@MOF

Equimolar amounts of ZrCl$_4$ and H$_2$BPDC were mixed in 50 mL DMF at 25°C to prepare UiO-66. After that, the formed solution was transferred to 100 mL autoclave (Teflon-lined stainless steel) and subjected to heat at 120°C for 16 hours. The product (UiO-66) was washed with the mixture of CH$_3$CH$_2$OH and distilled water and left to dry at ambient temperature. Then, it was heated to 70°C to eliminate excess reactants. The same procedure was applied to prepare NH$_2$-UiO-66 by using C$_8$H$_7$NO$_4$. Meanwhile, for the preparation of MIL-88 and NH$_2$-MIL-88, the metal source was changed to FeCl$_3$. To load imidazole (Im), 5 g of MOF sample in powder form was transferred to a mortar and 5 mL of imidazole solution was added drop-by-drop. The mixture was continuously ground and heated to 60°C for 12 hours to ensure that all imidazoles are completely impregnated inside the micropores of MOF.

13.4.3 X-RAY DIFFRACTION (XRD)

The crystallography properties of each samples were characterized using Panalytical Xpert3 Powder XRD with Cu Kα radiation (40 kV, 40 mA) at 2θ (2° to 80°) with scanning rate of 4°C min^{-1} .

13.4.4 SCANNING ELECTRON MICROSCOPY (SEM)

Evo LS15 Variable Pressure Scanning Electron Microscopy (VPSEM) was used to characterize the morphology of samples. The images of samples were captured at an acceleration voltage of 10 keV and an acceleration current of 10A.

13.4.5   Fourier Transformation Infrared Spectroscopy (FTIR)

Molecular characterizations of samples were done via Frontier 01 FTIR spectrometer by Perkin Elmer. Before the sample was analyzed, background spectrum was collected first to eliminate unwanted residue peak from the sample spectrum. The number of scans was set to 16 and the wavenumbers were recorded from 400 to 4000 cm^{-1}.

13.5   RESULTS AND DISCUSSION

13.5.1   X-Ray Diffraction

Figure 13.1 shows the diffractograms of MIL-88-Im, UiO-66- Im and their amino-functionalized forms. The major peaks of both MOFs were similar to the simulated XRD patterns, confirming the successful synthesis of these MOFs (Ma et al., 2013; Prabhu et al., 2021). Upon imidazole loading, the MIL-88-Im shows characteristic peaks at $2\theta = 6.95°$, $8.47°$, $21.24°$ and $23.7°$, while the main peaks were assigned to $7.27°$, $8.55°$, $21.26°$ and $23.55°$ in UiO-66-Im. The peaks were identical with the previous pristine MOF with slight shift of values, indicating that the structural integrity of MOF was preserved. By introducing amine group onto the organic linker of both MOFs, peak broadening was observed at $2\theta = 21°$, which can be attributed to smaller crystallite size (Figure 13.2). For UiO-66, the intensity of peaks at $7.27°$ and

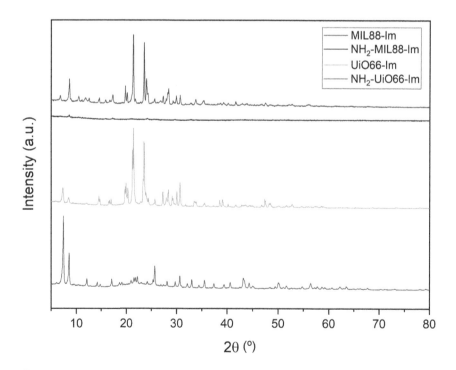

FIGURE 13.1   X-Ray diffractograms for each sample.

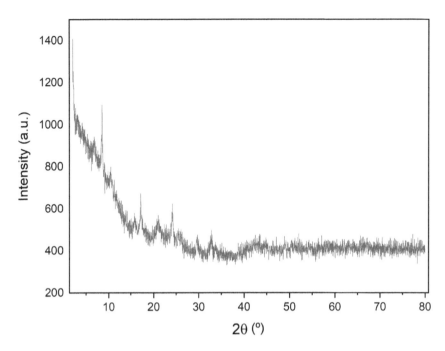

FIGURE 13.2   X-Ray diffractogram of NH$_2$-MIL-88-Im.

8.55° was increased, while it decreased at 21.26° and 23.55°, which suggests a more crystalline region.

13.5.2   MORPHOLOGY CHARACTERIZATIONS

Figures 13.3 and 13.4 show the morphology changes of MIL-88 and UiO-66 when impregnated with imidazole and upon functionalization of amine group. Tiny spindle-shape particles were observed in MIL-88 and its amino-functionalized form which was a typical pattern with previously reported images (Ma et al., 2013; Shi et al., 2015). Upon imidazole loading, well-ordered particles were observed, indicating that the framework structure was maintained without disturbance of guest molecules. Meanwhile, in UiO-66, irregular intergrown microstructure polyhedra were observed similar to the previous morphologies (Cao et al., 2015; Zhao et al., 2013; Zhao et al., 2019). Small cubic particles were also observed on the surface of MOF. When imidazole was incorporated, particles were agglomerated and leaving tiny void spaces. This feature might be useful for the transportation of Li-ions as it can be passed through the holes. Meanwhile, a fibrous-needle shape was observed in amino-functionalized MOF. The sharp and rough surface might be less suitable for solid electrolyte fabrication as lithium tends to grow on these regions, which can lead to short circuit (Porz et al., 2017). Smooth surface and less-defect surface are preferable and therefore some parameters need to be adjusted to control the morphology of the amino-functionalized MOF.

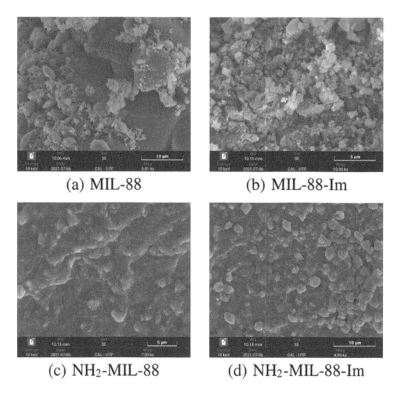

(a) MIL-88   (b) MIL-88-Im

(c) NH_2-MIL-88   (d) NH_2-MIL-88-Im

FIGURE 13.3   SEM images of MIL-88 and its functionalized forms.

13.5.3   FOURIER TRANSFORMATION INFRARED SPECTROSCOPY

The vibrational spectroscopy of MOFs and its amino-functionalized form were obtained from FTIR spectrum as shown in Figure 13.5. For all MOFs, characteristic peaks were observed around 1500–1600 cm^{-1} due to the symmetric and asymmetric carbonyl stretching from terephthalic acid. Primary amine was assigned at two peaks at 3456 cm^{-1} and 3310 cm^{-1} for NH2-MIL-88-Im, while for NH_2-UiO-66-Im, the peaks were located at 3431 cm^{-1} and 3300 cm^{-1} (Chakarova et al., 2019). These peaks were attributed to the symmetric and asymmetric stretching of N–H bond; however, the peaks were hard to observe due to overlap of tertiary amine from imidazole where it diminishes the amine peak. Due to the carbon stretching vibrations in imidazole ring, the peaks were observed at 1565 cm^{-1} and 1563 cm^{-1} for MIL-88 and UiO-66, respectively. The out-of-plane aromatic C–H stretching can also be observed around 750 cm^{-1}.

13.6   CONCLUSION

Two types of MOFs were successfully synthesized and functionalized with amino group and imidazole. Based on the XRD pattern, it can be observed that the MOFs had similar crystal structure with the pristine MOF and crystallinity was intensified

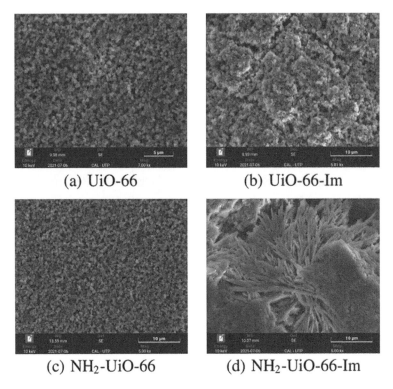

(a) UiO-66   (b) UiO-66-Im

(c) NH₂-UiO-66   (d) NH₂-UiO-66-Im

FIGURE 13.4   SEM images of UiO-66 and its functionalized forms.

when functionalized with amine and imidazole. The morphology of MIL-88 and UiO-66 exhibits well-ordered particles with tiny voids, which can facilitate the transportation of ions. However, a rough surface such as the one observed in NH2-UiO-66-Im was less preferable for SSE fabrication to avoid short circuit. Optimization of synthesis parameters is necessary to obtain a smooth surface. FTIR spectrums confirm the existence of important functional groups in the functionalized MOF. The molecular fingerprints also confirm the successful incorporation of imidazole onto the framework of MOFs.

ACKNOWLEDGMENT

This research is supported by Yayasan Universiti Teknologi PETRONAS (015LC0–283). The authors would like to thank Dr. Nadhratun Naiim Mobarak from the National University of Malaysia (UKM) for providing useful insights on solid-state electrolyte. Finally, the authors would like to thank Dr. Khairulazhar Jumbri, UTP Centre of Research in Ionic Liquids (CORIL), Centre of Innovative Nanostructures and Nanodevices (COINN) and UKM for providing the computational tools, laboratory facilities and analytical instruments throughout this research.

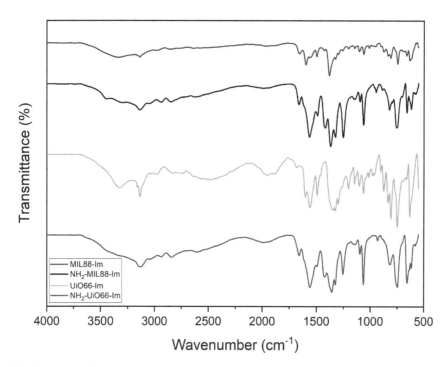

FIGURE 13.5   FTIR spectrums of all samples.

REFERENCES

Akhtar, S., & Ali, S. (2020). Characterization of nanomaterials: techniques and tools. In F. A. Khan (Ed.), *Applications of Nanomaterials in Human Health* (pp. 23–43). Springer, Singapore. https://doi.org/10.1007/978-981-15-4802-4_3

Al-Akraa, I. M., Mohammad, A. M., El-Deab, M. S., & El-Anadouli, B. E. (2017). Flower-shaped gold nanoparticles: preparation, characterization, and electrocatalytic application. *Arabian Journal of Chemistry*, *10*(6), 877–884. https://doi.org/10.1016/j.arabjc.2015.05.004

Anastas, P., & Eghbali, N. (2010). Green chemistry: principles and practice. *Chem. Soc. Rev.*, *39*(1), 301–312. https://doi.org/10.1039/B918763B

Arbabzadeh, M., Sioshansi, R., Johnson, J. X., & Keoleian, G. A. (2019). The role of energy storage in deep decarbonization of electricity production. *Nature Communications*, *10*(1), 3413. https://doi.org/10.1038/s41467-019-11161-5

Aziz, S. B., Woo, T. J., Kadir, M. F. Z., & Ahmed, H. M. (2018). A conceptual review on polymer electrolytes and ion transport models. *Journal of Science: Advanced Materials and Devices*, *3*(1), 1–17.

Baumann, A. E., Burns, D. A., Liu, B., & Thoi, V. S. (2019). Metal-organic framework functionalization and design strategies for advanced electrochemical energy storage devices. *Communications Chemistry*, *2*(1), 1–14.

Cao, Y., Zhao, Y., Lv, Z., Song, F., & Zhong, Q. (2015). Preparation and enhanced CO_2 adsorption capacity of UiO-66/graphene oxide composites. *Journal of Industrial and Engineering Chemistry*, *27*, 102–107. https://doi.org/10.1016/j.jiec.2014.12.021

Chakarova, K., Strauss, I., Mihaylov, M., Drenchev, N., & Hadjiivanov, K. (2019). Evolution of acid and basic sites in UiO-66 and UiO-66-NH2 metal-organic frameworks: FTIR study by probe molecules. Microporous and Mesoporous Materials, 281, 110–122. https://doi. org/10.1016/j.micromeso.2019.03.006

Chatterjee, K., Pathak, A. D., Lakma, A., Sharma, C. S., Sahu, K. K., & Singh, A. K. (2020). Synthesis, characterization and application of a non-flammable dicationic ionic liquid in lithium-ion battery as electrolyte additive. Scientific Reports, 10(1), 1–12.

Chen, T.-L., Kim, H., Pan, S.-Y., Tseng, P.-C., Lin, Y.-P., & Chiang, P.-C. (2020). Implementation of green chemistry principles in circular economy system towards sustainable development goals: challenges and perspectives. Science of the Total Environment, 716, 136998. https://doi.org/10.1016/j.scitotenv.2020.136998

Dugershaw, B. B., Aengenheister, L., Hansen, S. S. K., Hougaard, K. S., & Buerki-Thurnherr, T. (2020). Recent insights on indirect mechanisms in developmental toxicity of nanomaterials. Particle and Fibre Toxicology, 17(1), 31. https://doi.org/10.1186/s12 989-020-00359-x

Fujie, K., Yamada, T., Ikeda, R., & Kitagawa, H. (2014). Introduction of an ionic liquid into the micropores of a metal–organic framework and its anomalous phase behavior. Angewandte Chemie, 126(42), 11484–11487.

Ganguly, P., Breen, A., & Pillai, S. C. (2018). Toxicity of nanomaterials: exposure, pathways, assessment, and recent advances. ACS Biomaterials Science & Engineering, 4(7), 2237–2275. https://doi.org/10.1021/acsbiomaterials.8b00068

Hallan, S. S., Sguizzato, M., Esposito, E., & Cortesi, R. (2021). Challenges in the physical characterization of lipid nanoparticles. Pharmaceutics, 13(4), 549. https://doi.org/10.3390/pharmaceutics13040549

Irshad, M. A., Nawaz, R., Rehman, M. Z. ur, Adrees, M., Rizwan, M., Ali, S., Ahmad, S., & Tasleem, S. (2021). Synthesis, characterization and advanced sustainable applications of titanium dioxide nanoparticles: a review. Ecotoxicology and Environmental Safety, 212, 111978. https://doi.org/10.1016/j.ecoenv.2021.111978

Jaramillo-Cabanzo, D. F., Ajayi, B. P., Meduri, P., & Sunkara, M. K. (2021). One-dimensional nanomaterials in lithium-ion batteries. Journal of Physics D: Applied Physics, 54(8), 083001. https://doi.org/10.1088/1361-6463/abc3eb

Kim, J., Kwon, J., Kim, M. J., O, M. J., Jung, D. S., Roh, K. C., Jang, J., Kim, P. J., & Choi, J. (2021). A strategic approach to use upcycled Si nanomaterials for stable operation of lithium-ion batteries. Nanomaterials, 11(12), 3248. https://doi.org/10.3390/nano1 1123248

Kumar, N., & Kumbhat, S. (2016). Essentials in Nanoscience and Nanotechnology (1st ed.). Wiley, United States. https://doi.org/10.1002/9781119096122

Li, Y., Lv, W., Huang, H., Yan, W., Li, X., Ning, P., Cao, H., & Sun, Z. (2021). Recycling of spent lithium-ion batteries in view of green chemistry. Green Chemistry, 23(17), 6139–6171. https://doi.org/10.1039/D1GC01639C

Liang, S., Yan, W., Wu, X., Zhang, Y., Zhu, Y., Wang, H., & Wu, Y. (2018). Gel polymer electrolytes for lithium ion batteries: fabrication, characterization and performance. Solid State Ionics, 318, 2–18.

Lin, L., Peng, H., & Liu, Z. (2019). Synthesis challenges for graphene industry. Nature Materials, 18(6), 520–524. https://doi.org/10.1038/s41563-019-0341-4

Liu, S., Yue, Z., & Liu, Y. (2015). Incorporation of imidazole within the metal–organic framework UiO-67 for enhanced anhydrous proton conductivity. Dalton Transactions, 44(29), 12976–12980. https://doi.org/10.1039/c5dt01667c

Ma, M., Noei, H., Mienert, B., Niesel, J., Bill, E., Muhler, M., Fischer, R. A., Wang, Y., Schatzschneider, U., & Metzler-Nolte, N. (2013). Iron metal-organic frameworks

MIL-88B and NH_2-MIL-88B for the loading and delivery of the gasotransmitter carbon monoxide. *Chemistry—A European Journal, 19*(21), 6785–6790. https://doi.org/10.1002/chem.201201743

Megías-Sayago, C., Bingre, R., Huang, L., Lutzweiler, G., Wang, Q., & Louis, B. (2019). CO_2 adsorption capacities in zeolites and layered double hydroxide materials. *Frontiers in Chemistry, 7,* 551. https://doi.org/10.3389/fchem.2019.00551

Mitchell, M. J., Billingsley, M. M., Haley, R. M., Wechsler, M. E., Peppas, N. A., & Langer, R. (2021). Engineering precision nanoparticles for drug delivery. *Nature Reviews Drug Discovery, 20*(2), 101–124. https://doi.org/10.1038/s41573-020-0090-8

Moi, R., Ghorai, A., Banerjee, S., & Biradha, K. (2020). Amino- and sulfonate-functionalized metal–organic framework for fabrication of proton exchange membranes with improved proton conductivity. *Crystal Growth & Design, 20*(8), 5557–5563. https://doi.org/10.1021/acs.cgd.0c00732

Nagpure, S. C., & Bhushan, B. (2016). Nanomaterials for electrical energy storage devices. In B. Bhushan (Ed.), *Encyclopedia of Nanotechnology* (pp. 2473–2485). Springer, Netherlands. https://doi.org/10.1007/978-94-017-9780-1_58

Omran, B. A., & Baek, K.-H. (2022). Valorization of agro-industrial biowaste to green nanomaterials for wastewater treatment: approaching green chemistry and circular economy principles. *Journal of Environmental Management, 311,* 114806. https://doi.org/10.1016/j.jenvman.2022.114806

Poerwoprajitno, A. R., Gloag, L., Watt, J., Cychy, S., Cheong, S., Kumar, P. V., Benedetti, T. M., Deng, C., Wu, K., Marjo, C. E., Huber, D. L., Muhler, M., Gooding, J. J., Schuhmann, W., Wang, D., & Tilley, R. D. (2020). Faceted branched nickel nanoparticles with tunable branch length for high-activity electrocatalytic oxidation of biomass. *Angewandte Chemie International Edition, 59*(36), 15487–15491. https://doi.org/10.1002/anie.202005489

Porz, L., Swamy, T., Sheldon, B. W., Rettenwander, D., Frömling, T., Thaman, H. L., Berendts, S., Uecker, R., Carter, W. C., & Chiang, Y. (2017). Mechanism of lithium metal penetration through inorganic solid electrolytes. *Advanced Energy Materials, 7*(20), 1701003. https://doi.org/10.1002/aenm.201701003

Prabhu, S. M., Chuaicham, C., Park, C. M., Jeon, B.-H., & Sasaki, K. (2021). Synthesis and characterization of defective UiO-66 for efficient co-immobilization of arsenate and fluoride from single/binary solutions. *Environmental Pollution, 278,* 116841. https://doi.org/10.1016/j.envpol.2021.116841

Remya, R. R., Julius, A., Suman, T. Y., Mohanavel, V., Karthick, A., Pazhanimuthu, C., Samrot, A. V., & Muhibbullah, M. (2022). Role of nanoparticles in biodegradation and their importance in environmental and biomedical applications. *Journal of Nanomaterials, 2022,* 1–15. https://doi.org/10.1155/2022/6090846

Rhodes, D., Chae, S. H., Ribeiro-Palau, R., & Hone, J. (2019). Disorder in van der Waals heterostructures of 2D materials. *Nature Materials, 18*(6), 541–549. https://doi.org/10.1038/s41563-019-0366-8

Rogers, L., & Jensen, K. F. (2019). Continuous manufacturing—the green chemistry promise? *Green Chemistry, 21*(13), 3481–3498. https://doi.org/10.1039/C9GC00773C

Roth, E. P., & Orendorff, C. J. (2012). How electrolytes influence battery safety. *Electrochemical Society Interface, 21*(2), 45.

Sadiq, M. M., Konstas, K., Falcaro, P., Hill, A. J., Suzuki, K., & Hill, M. R. (2020). Engineered porous nanocomposites that deliver remarkably low carbon capture energy costs. *Cell Reports Physical Science, 1*(6), 100070. https://doi.org/10.1016/j.xcrp.2020.100070

Salame, P. H., Pawade, V. B., & Bhanvase, B. A. (2018). Characterization tools and techniques for nanomaterials. In Dhoble, S. J., Sonawane, S. H. & Ashokkumar, M. (eds),

Nanomaterials for Green Energy (pp. 83–111). Elsevier, Netherlands. https://doi.org/10.1016/B978-0-12-813731-4.00003-5

Scarlat, N., Dallemand, J.-F., & Fahl, F. (2018). Biogas: developments and perspectives in Europe. *Renewable Energy, 129*, 457–472. https://doi.org/10.1016/j.renene.2018.03.006

Scarlat, N., Dallemand, J.-F., Monforti-Ferrario, F., Banja, M., & Motola, V. (2015). Renewable energy policy framework and bioenergy contribution in the European Union—an overview from National Renewable Energy Action Plans and Progress Reports. *Renewable and Sustainable Energy Reviews, 51*, 969–985. https://doi.org/10.1016/j.rser.2015.06.062

Shi, L., Wang, T., Zhang, H., Chang, K., Meng, X., Liu, H., & Ye, J. (2015). An Amine-functionalized iron(III) metal-organic framework as efficient visible-light photocatalyst for Cr(VI) reduction. *Advanced Science, 2*(3), 1500006. https://doi.org/10.1002/advs.201500006

Singh, P., Kim, Y. J., Wang, C., Mathiyalagan, R., & Yang, D. C. (2016). Microbial synthesis of flower-shaped gold nanoparticles. *Artificial Cells, Nanomedicine, and Biotechnology, 44*(6), 1469–1474. https://doi.org/10.3109/21691401.2015.1041640

Singh, A., Vedarajan, R., & Matsumi, N. (2017). Modified metal organic frameworks (MOFs)/ionic liquid matrices for efficient charge storage. *Journal of the Electrochemical Society, 164*(8), H5169–H5174.

Sneddon, G., Greenaway, A., & Yiu, H. H. P. (2014). The potential applications of nanoporous materials for the adsorption, separation, and catalytic conversion of carbon dioxide. *Advanced Energy Materials, 4*(10), 1301873. https://doi.org/10.1002/aenm.201301873

Sreedharan, S. M., Singh, S. P., & Singh, R. (2019). Flower shaped gold nanoparticles: biogenic synthesis strategies and characterization. *Indian Journal of Microbiology, 59*(3), 321–327. https://doi.org/10.1007/s12088-019-00804-2

Su, S., & Kang, P. M. (2020). Systemic review of biodegradable nanomaterials in nanomedicine. *Nanomaterials, 10*(4), 656. https://doi.org/10.3390/nano10040656

Tambelli, C. C., Bloise, A. C., Rosario, A. V., Pereira, E. C., Magon, C. J., & Donoso, J. P. (2002). Characterisation of PEO–Al$_2$O$_3$ composite polymer electrolytes. *Electrochimica Acta, 47*(11), 1677–1682.

Thompson, D. L., Hartley, J. M., Lambert, S. M., Shiref, M., Harper, G. D. J., Kendrick, E., Anderson, P., Ryder, K. S., Gaines, L., & Abbott, A. P. (2020). The importance of design in lithium ion battery recycling—a critical review. *Green Chemistry, 22*(22), 7585–7603. https://doi.org/10.1039/D0GC02745F

van der Merwe, D., Pickrell, J. A. (2018). Toxicity of nanomaterials. In Gupta, R. C. (ed.), *Veterinary Toxicology* (pp. 319–326). Elsevier, Netherlands. https://doi.org/10.1016/B978-0-12-811410-0.00018-0

Vishwakarma, V., Samal, S. S., & Manoharan, N. (2010). Safety and risk associated with nanoparticles—a review. *Journal of Minerals and Materials Characterization and Engineering, 09*(05), 455–459. https://doi.org/10.4236/jmmce.2010.95031

Wang, M., Chen, T., Liao, T., Zhang, X., Zhu, B., Tang, H., & Dai, C. (2021). Tin dioxide-based nanomaterials as anodes for lithium-ion batteries. *RSC Advances, 11*(2), 1200–1221. https://doi.org/10.1039/D0RA10194J

Wang, S., Luo, H., Li, X., Shi, L., Cheng, B., Zhuang, X., & Li, Z. (2021). Amino acid-functionalized metal organic framework with excellent proton conductivity for proton exchange membranes. *International Journal of Hydrogen Energy, 46*(1), 1163–1173. https://doi.org/10.1016/j.ijhydene.2020.09.235

Xu, Q., Zhang, X., Zeng, S., Bai, L., & Zhang, S. (2019). Ionic liquid incorporated metal organic framework for high ionic conductivity over extended temperature range. *ACS Sustainable Chemistry & Engineering, 7*(8), 7892–7899.

Yaroslavtsev, A. B., Kulova, T. L., & Skundin, A. M. (2015). Electrode nanomaterials for lithium-ion batteries. *Russian Chemical Reviews*, *84*(8), 826–852. https://doi.org/10.1070/RCR4497

Yoshida, Y., & Kitagawa, H. (2018). Ionic conduction in metal–organic frameworks with incorporated ionic liquids. *ACS Sustainable Chemistry & Engineering*, *7*(1), 70–81.

Zagho, M. M., Hassan, M. K., Khraisheh, M., Al-Maadeed, M. A. A., & Nazarenko, S. (2021). A review on recent advances in CO_2 separation using zeolite and zeolite-like materials as adsorbents and fillers in mixed matrix membranes (MMMs). *Chemical Engineering Journal Advances*, *6*, 100091. https://doi.org/10.1016/j.ceja.2021.100091

Zhao, R., Wu, Y., Liang, Z., Gao, L., Xia, W., Zhao, Y., & Zou, R. (2020). Metal-organic frameworks for solid-state electrolytes. *Energy & Environmental Science*, *13*(8), 2386–2403.

Zhao, X., Yuan, Y., Li, P., Song, Z., Ma, C., Pan, D., Wu, S., Ding, T., Guo, Z., & Wang, N. (2019). A polyether amine modified metal organic framework enhanced the CO_2 adsorption capacity of room temperature porous liquids. *Chemical Communications*, *55*(87), 13179–13182. https://doi.org/10.1039/C9CC07243H

Zhao, Q., Yuan, W., Liang, J., & Li, J. (2013). Synthesis and hydrogen storage studies of metal–organic framework UiO-66. *International Journal of Hydrogen Energy*, *38*(29), 13104–13109. https://doi.org/10.1016/j.ijhydene.2013.01.163

Zimmerman, J. B., Anastas, P. T., Erythropel, H. C., & Leitner, W. (2020). Designing for a green chemistry future. *Science*, *367*(6476), 397–400. https://doi.org/10.1126/science.aay3060

14 Green Removal of Bisphenol A from Aqueous Media Using Zr-Based Metal-Organic Frameworks

Afzan Mahmad,[1] *Teh Ubaidah Noh,*[3] *and Maizatul Shima Shaharun*[2*]

[1,2] Fundamental and Applied Sciences Department, Universiti Teknologi PETRONAS, Bandar Seri Iskandar, 32610 Seri Iskandar, Perak Darul Ridzuan, Malaysia

[3] Institute of Bioproduct Development, Universiti Teknologi Malaysia

* Corresponding author: Maizatul Shima Shaharun, maizats@utp.edu.my

CONTENTS

14.1 INTRODUCTION

Water pollution due to toxic organic and inorganic compounds has been a common phenomenon affecting the harmony of an ecosystem. Endocrine–disrupting compounds such as Bisphenol A (BPA) have been frequently detected in industrial and environmental waters [1]. BPA is commonly utilised in the production of storage units such as plastic products [1], groundwater pipes [2], plastic bottles

DOI: 10.1201/9781003320746-14

[3–5], and food containers [6, 7]. With two phenol moieties [8], BPA has low vapour pressure, moderate water solubility, and low volatility. Aquatic organisms are acutely toxic to 1,000–10,000 mg/L for marine environments [9, 10]. The BPA, on the other hand, has been shown to exhibit estrogenic action at doses lower than 1 mg/m^3 [11].

In order to remove BPA from wastewater, many wastewater treatment methods have been developed to date [9, 12]. Adsorption has been widely accepted as one of the promising techniques to remove organic pollutants [8]. The most commonly used adsorbent is activated carbon at the laboratory and industrial scale. However, it is plagued with disadvantages, such as problems with hydrophilic substances, the limited performance that depends on the type of carbon used, the required complexing agents to enhance their efficiency of removal, and not being effective in the adsorption removal of low concentrations of pollutants [2]. Thus, alternative adsorbents must be introduced to treat the BPA wastewater so that the discharged water can comply with environmental regulations.

Metal-organic frameworks (MOFs) are suggested as one of the most advanced materials that can effectively remove toxic pollutants from wastewater. MOFs are crystalline porous organic–inorganic hybrids that feature positively charged metal ions adjacent to organic linker molecules. Besides, many advantages ranging from an increased outstanding [13, 14] and ultra–porous structure for the surface area [15], pore volumes [16], uniform particle sizes [17–20], and flexibility of geometrical structures to impart different shapes for the frameworks [9] have been reported. These advantages of MOFs have the potential for adsorption approaches for the removal of BPA in wastewater. UiO–66(Zr) MOF is one of the future adsorbents that have the possible reusability for rapid adsorption of BPA in wastewater. This review will preview the possibility of employing UiO–66(Zr) MOF as an adsorbent to remove the BPA in wastewater.

14.2   BISPHENOL A IN WASTEWATER

BPA manufacture originated in the United States in 1957 and in Europe in 1958. In 2004, the average output of BPA in the United States was expected to be one million tonnes (2.3 billion pounds), while in 2005, Western Europe generated more than one million tonnes [21]. Annually, the growth of global production of BPA has been estimated to be in the range of 5%, with China being reported to be the latest strongest growth development of BPA users [22]. Throughout 2000 to 2006, Asia's BPA market grew at an annual rate of 13% on average [22]. According to these findings, BPA is a demand–producing substance in the United States, Europe, and Asia. BPA, as previously noted, is an endocrine–disrupting molecule for flora and fauna that is widely employed in the industry. The chemical functions as a monomer, which is good for making epoxy resins and polycarbonate polymers. The concentration of BPA varies depending on temperature, pH, source, and sample period in different sites [23]. Most of the BPA released by the industry ends up in landfills in which the compound leached into water and paper recycling. Some municipal wastewater has contained BPA within low concentrations [24].

BPA has permeated the environment as a result of industrial or home sewage effluent discharge. BPA concentrations in wastewater, sewage samples, leachate, and drinkable or natural mineral water are been shown to be the highest. BPA contamination is most prevalent in rivers and lakes. BPA compounds can be discovered in factory effluent due to improper removal and endangering aquatic life during the primary treatment [25]. Yamamoto et al. [26] discovered that landfill leachate has a high amount of BPA (1.3–17,200 g/L), which is most likely owing to it leaking out from plastic trash. Suzuki et al. [27] determined that the BPA level in the effluent is greater because it is discharged straight into the environment without treatment. However, Huang et al. [22] discovered in another study that BPA exposure is greater in commercial and industrial areas. BPA traces released from wastewater treatment have a high probability of collecting in sewage sludge.

Besides that, BPA concentrations in the wastewater treatment plant (WWTP) effluent were reported to be in the range from non–detected to 370 mg/L. In surface water, the estimated BPA is to be from non–detected to 56 mg/L. In addition, river sites upstream of the WWTP had their surface water samples assessed, and BPA was analysed in marine and coastal systems [28]. The highest exceedance of effluent was recorded in Canada followed by North America and Asia with 63.4%, 56.3%, and 52.4%, respectively. The BPA exceedance in surface water from Asia exceeded the Canadian 80% of the time, while North America and Europe exceedances were observed at 53.1% and 34.6%, respectively [29]. These data suggested that Europe and North America have the highest percentage of exceedance in affluent compared to the surface water [30].

In Malaysia, the BPA concentration is reported in tap water (0.0035–0.0598 µg/L) and bottled mineral water (0.0007–0.0059 µg/L) [23]. To date, Malaysia has not established any official standards regarding the safe level of BPA in wastewater discharge. However, most developing countries such as the United States and Canada have created awareness on the safety of using water to consumers through their environmental organisations. Also, the awareness of BPA in packaging products has gained attention from the consumer. There is still a lack of attention by the Malaysian government regarding the pollution of wastewater sources, especially at the BPA level. Except in Taiwan, BPA concentrations in rivers were less than 0.01 g/L. Furthermore, France has the highest BPA levels in bottled water (0.07–4.21 g/L). This is due to BPA produced from polycarbonate drinking bottles when exposed to elevated temperatures [31].

14.3   METHODS FOR THE BPA REMOVAL IN WATER

BPA removal from water has been accomplished using a variety of approaches. Adsorption is the most common method cited by many researchers [32]. There are also other methods such as nanofiltration, advanced oxidation, reverse osmosis (membrane separation), ultrafiltration membranes, and biodegradation process that show promising performance in the removal of BPA as shown in Table 14.1. These data were analysed using different approaches based on the removal methods of BPA in the wastewater.

TABLE 14.1
Various Methods Reported for the Removal of BPA

Method	Adsorbent	Adsorption capacity	Percentage Removal (%)	Isotherm Model	Kinetics Model	References
Nanofiltration and reverse osmosis membranes	Membranes NF 90, NF 270, XLE BWRO, BW 30 (Dow FilmTech), CE BWRO and AD SWRO (GE Osmonics)	–	≥98	–	–	[33]
Advanced oxidation and reverse osmosis (membrane separation)	UV/H_2O_2	–	60–80	–	–	[34]
Ultrafiltration membranes	Natural organic matter and single–walled carbon nanotubes	$4.21\,\mu g/cm^2$	–	Langmuir	–	[35]
Biodegradation process	*Serratia rubidiae*, *Pseudomonas aeruginosa* and *Escherichia coli* K12	–	>90	–	–	[36]
Purification catalysis	$TAML/H_2O_2$ catalysis readily oxidises BPA	–	>99	–	–	[37]
	Nano zero–valent iron/palygorskite	–	99	–	–	[38]
Adsorption	Activated carbons	$1641\ m^2g/l$	–	Langmuir	–	[39]
	Organo Arizona SAz–2 Ca montmorillonite	151.52 mg/g	–	Langmuir	Pseudo–second order	[40]
	Graphene	–	–	Freundlich	Pseudo–second order	[41]
	Organo–montmorillonites	222.2 mg/g	–	Langmuir	Pseudo–second order	[42]
	Goethite iron oxide particles impregnated activated carbon composite	$56.60\,\mu g/cm^1$	–	Freundlich	Pseudo–second order	[43]

Moreira et al. reported a study on the removal of BPA using nanofiltration and reverse osmosis membranes methods. The NF 90, XLE BWRO, CE BWRO, NF 270, AD SWRO (GE Osmonics), and BW 30 (Dow FilmTech) membranes were studied by the researchers. Polyamide–based membranes outperformed the cellulose acetate–based membranes in terms of BPA removal efficacy [33]. Moreira et al. [44] reported the removal of BPA using two treatment processes: UV/H_2O_2 advanced oxidation and reverse osmosis (membrane separation). The UV/H_2O_2 performance showed 48% efficiency to remove the BPA compound [44].

Another study by Heo et al. [35] studied the adsorption of BPA using ultrafiltration membranes. A stirred cell operated within a batch dead–end stirred cell was determined in single–walled carbon nanotubes and natural organic matter. The NOM fouling represents the increased adsorptive hydrophobic interactions in which low membrane pore size was used in the removal of BPA [35]. Mita et al. [36] analysed *Pseudomonas aeruginosa*, *Serratia rubidiae*, and *Escherichia coli* for BPA removal and biofilm formation on the activated granule carbon in the biodegradation process. *P. aeruginosa* was found as the best strain to be employed in the BPA removal process. The yield in the BPA removal of a *P. aeruginosa* biofilm grown was increased [36].

Onundi et al. [37] reported that tetraamidomacrocyclic ligand (TAML) by H_2O_2 catalysis oxidises BPA and the tetrachloro, ring–tetramethyl, and tetrabromo–substituted derivatives. The TAML/H_2O_2 oxidations with second–order rates were used for the substrate oxidation process when the pH was set to 11. TAML/H_2O_2 catalysis removed BPA with relative ease and high efficiency, suggesting a straightforward and innovative technique for BPA removal [37]. The batch adsorption of BPA removal was analysed by Zheng et al. [34] using various adsorbents. With varied surfactant (hexadecyltrimethyl ammonium (HDTMA) and Dodecyltrimethylammonium bromide (DDTMA)) loadings, the organo Arizona SAz–2 Ca–montmorillonite was measured by direct ion exchange (DIE). The adsorption of BPA was accomplished by the formation of a hydrophobic phase and a positively charged surface by the loading of surfactant molecules. The organoclays that have been adsorbed on the surface containing longer chain surfactant molecules have a greater adsorption capacity for BPA in alkaline environments than conventional organoclays. Using this approach, BPA may be appealingly removed from polluted water [40].

The removal of BPA on activated carbons has been investigated [39]. Olive–mill waste activated carbon demonstrated enhanced surface area (up to 1641 m^2g/L) and a significant heterogeneous micropore distribution as compared to other activated carbons. The enhanced adsorption of BPA is associated with the *in situ* synthesis of complex compounds in which activated carbons serve as the core metallic cation [39]. Xi et al. [38] described the usage of a nano zero–valent iron/palygorskite compound for the removal efficiency of BPA in aqueous solutions using centrifugal and evaporative techniques in a recent publication. By using hydrogen peroxide and air bubbles, the proportion of BPA removed improved dramatically, reaching 99% [38].

In a separate publication, Kwon et al. [41] investigated the removal efficiency of BPA by graphene in an aqueous solution. Graphene is made up of carbon atoms that are arranged in layers and held together by van der Waals forces, due to increased current density and surface area. In this study, reduced graphene oxide (rGO) was produced by employing two different methods: the chemical reduction approach (H–rGO) and the thermal exfoliation method (T–rGO). In comparison to BPA, T–rGo demonstrated at least a 2.5–fold increase in adsorption capabilities. Furthermore, T–rGO exhibited a kinetic model constant that was 200 times greater than that of H–rGO. This finding demonstrates that graphene is efficient for the elimination of BPA when using the adsorption approach [41]. In 2016, Yang et al. [42] reported the interaction and the hydrophobic affinity of organo–montorillonites modified with a long alkyl chain for the effective removal of BPA. The results showed that the maximum adsorption capacities of BPA were obtained using organoclays with shorter alkyl chains [43]. Additionally, batch adsorption tests were conducted to determine the adsorption properties of goethite iron oxide particles impregnated with an activated carbon composite containing BPA. It was also shown that the adsorption capability of BPA increased as the quantity of iron oxide in the solution increased [43].

14.4 ADSORPTION OF BISPHENOL A IN WASTEWATER

Adsorption is an attractive alternative method when the adsorbent is readily available and is at a low cost [45, 46]. Adsorption is most famously cited for the removal of inorganic and organic compounds due to ease of design and operation. Besides low–cost operations, the complete adsorption of the pollutants is achieved within a short time and is environmentally friendly [47]. Adsorption is a surface phenomenon that involves a mixture of direct pore condensation and surface binding of the adsorbate. Pore condensation is crucial for aqueous near or below boiling point and adsorbate direct binding dominates at higher temperatures [48]. It is important to distinguish between two fundamentally different kinds of adsorption, which are multilayer adsorption whereby in both multilayer and monolayer adsorption, the adsorbate is held near to the solid surface, and the adsorbate is retained close to the solid surface [49, 50].

14.5 ADSORBENTS REPORTED FOR THE REMOVAL OF BPA FROM WATER

Adsorbents can be classified into five types: (i) natural materials (e.g., sawdust, fuller's earth, bauxite, or wood); (ii) natural materials that were modified (e.g., silica gel, activated carbons, or activated alumina); (iii) manufactured materials (e.g., zeolites or polymeric resins); (iv) industrial by–products and agricultural solid wastes (e.g., red mud, date pits or fly ash) and (v) biosorbents (e.g., fungi, chitosan or bacterial biomass) [30].

Activated carbon is sometimes mixed with other materials to form composites that were tested for the adsorption of BPA. Noufel et al. [51] reported three adsorbents by the encapsulation of activated carbon, organo–activated bentonite, and activated carbon/organo–activated bentonite in cross–linked alginate beads for the removal of BPA. The activated carbon/alginate beads reported the maximum BPA adsorption capacities with 444.7 and 419.3 mg/g, respectively. Zhou et al. [52] reported the preparation of powdered activated carbon embedded chitosan–polyvinyl alcohol to produce a composite (PAC/CS/PVA) bead. The authors found that increased adsorbent dosages produced increased removal efficiencies but lower removal capacities. Mohanta et al. [53] studied the adsorption and decontamination of BPA through the coupled adsorption–photodegradation process. A biogenic synthetic route for the fabrication of SnO_2 quantum dot encapsulated carbon nanoflake (SnO_2–CNF) as a combination of photocatalytic adsorbent has been carried out. Hydrogen bonding and π–π interaction are possible for the significant removal of BPA with SnO_2–CNF [53].

Furthermore, Liu et al. studied the preparation of Fe_3O_4@Co/Ni–LDH for the removal of BPA. The adsorbent was characterised by a hierarchical rattle–like structure and possesses an abundant pore system, high specific surface area, and magnetic properties. The removal of BPA was primarily due to the hydrogen bond interaction between Fe_3O_4@Co/Ni–LDH and BPA [54]. According to Cao et al. [55], photocatalytically active Bi_5O_7I has been combined with UiO–67 to form Bi_5O_7I/ UiO–67 heterojunctions. The results revealed that with a ratio of 4:1, Bi_5O_7I /UiO–67 was excellent in decomposing BPA within an hour of being added to the solution. Sadeghzadeh et al. disclosed the synthesis of functionalised magnetic Fe_3O_4 nanoparticles that were changed by surface modification using (3–aminopropyl) trimethoxysilane using the co–precipitation process, which he discussed in detail. Using a starting concentration of 60 ppm, the highest removal effectiveness of 87.3% was achieved during 11 hours of BPA removal [56]. The use of polymer resin was also reported for the removal adsorption of BPA. Katančić et al. [57] analysed the composite photocatalyst made of TiO_2 and conductive polymer poly (3,4–ethylenedioxythiophene) (TiO_2–PEDOT) for the adsorption of BPA. TiO_2–PEDOT with different PEDOT contents were determined by oxidative chemical polymerisation of 3,4–ethylenedioxythiophene (EDOT) in the presence of TiO_2 using either $FeCl_3$ or $(NH_4)_2S_2O_8$ as oxidants to analyse the photocatalytic properties. Based on the results of the study, scavenging bulk and surface hydroxyl radicals effectively remove BPA from water. Table 14.2 presents the summary of various adsorbents reported for the removal of BPA.

14.6 METAL-ORGANIC FRAMEWORK (MOF)

The characteristics of MOF show suitable adsorbents for the adsorption of BPA. The MOF as a functional material has great characteristics, including a tuneable ultra–high porosity, crystalline nature, changeable structure, versatile functionality, large pore volumes and surface area, uniform pore size, specific interaction sites for an excellent

TABLE 14.2
Various Adsorbents Reported for the Removal of BPA

Type of Adsorbents	Adsorbents in the Work	Disadvantage	Adsorption Capacity (mg/g)	Percentage Removal (%)	Isotherm Model	Kinetics Model	References
Activated carbon	Organo–activated bentonite, activated carbon, and organo–activated bentonite/activated carbon	The higher the quality, the greater the cost.	419.3	–	Langmuir Freundlich	Pseudo–second order	[51]
	SnO$_2$ quantum dot encapsulated carbon nanoflake (SnO$_2$–CNF) as an integrated photocatalytic adsorbent	Performance is dependent on the type of carbon used. A non-selective process	250	–	Langmuir	Pseudo–second order	[53]
	Activated carbon embedded chitosan–polyvinyl	Problems with hydrophilic substances	–	73.4	–	–	[52]
MOFs	E. coli@UiO–67 composites	Commercial MOFs are quite expensive	402.930	–	Langmuir Freundlich	Pseudo–second order	[57]
	Fe$_3$O$_4$@Co/Ni–LDH		–	2.0–4.5	–	–	[54]
	UiO–67 with Bi$_5$O$_7$I/UiO–67 heterojunctions		–	100	–	Pseudo–first order	[55]
Polymer	Amino–functionalized magnetic Fe$_3$O$_4$	Commercial resins are quite expensive. Performance is dependent on the type of resin used	–	57.3	–	–	[58]
	TiO$_2$ and conductive polymer poly(3,4–ethylenedioxythiophene) (TiO$_2$–PEDOT)		–	20	–	–	[57]

optoelectronic property, mechanically stable, and highly thermal adsorbate [59]. MOF was prepared with porosity and ultra–high surface area that was well arranged due to the crystalline nature. Furthermore, *in situ* structural and post–synthetic alterations to MOF are effective for achieving extraordinary material properties while not changing the fundamental network topology for BPA adsorption [60].

Surface area and porosity are major factors influencing an adsorbent's adsorption capability. Normal adsorption–desorption isotherms for nitrogen (N_2) are used to estimate the surface area and pore characteristics of a material. To estimate the surface area, the Brunaeur, Emmett, and Teller (BET) approach might be utilised. This method determines the amount of nitrogen adsorbed onto or desorbed from a test material at 77 K. The total pore volume is calculated by multiplying the gas adsorbed by 0.98.

In general, a large surface area identifies the existence of substantial porosity [61, 62]. Pore size is divided into three types: micropores (diameter ≤ 2 nm), mesopores (diameter 2–50nm), and macropores (diameter ≥ 50 nm). The adsorbent's adsorptive capabilities are controlled by the pore size [41]. Small pore sizes are not capable of trapping big adsorbate molecules, whereas large pore sizes are incapable of trapping small adsorbate molecules, whether charged, polar molecules, or uncharged non–polar chemicals [42]. With the stated advantages, UiO–66 (Zr) MOF was proposed as a future adsorbent MOF to study the removal batch adsorption of BPA.

14.7 UIO–66 (ZR) METAL-ORGANIC COMPOUND IN BPA REMOVAL IN WASTEWATER

Recently, there have been attempts to synthesise MOF using relatively simple procedures and inexpensive chemicals. It has been demonstrated that microwave–assisted techniques, solvent–free routes, and the use of MOF precursors derived from waste materials may all be used [16]. To make Zr–based MOF (UiO–66 (Zr)), it is common to employ terephthalic acid (H_2BDC) manufactured using waste polyethylene terephthalate (PET) as a starting material. MOFs are important subjects for industrial characteristics such as high surface areas and micropore volumes, which make them ideal for physisorption–based hydrogen preservation and environmentally friendly manufacturing processes [63, 64].

UiO–66 (Zr) is zirconium–based MOF with a greater surface area and stability (1180–1240 m^2/g). UiO–66 (Zr) is made up of Zr_6O_4 (OH)$_4$ octahedra that are 12–fold coupled to the neighbouring octahedra through a 1,4–benzenedicarboxylate (BDC) linker (Figure 14.1). Zr–O linkages generated between carboxylate ligands and clusters have been found to boost the stability of Zr–based MOF. Furthermore, the ability of the core Zr6–cluster and strong ZrO bonds was rearranged reversibly upon addition or removal of 3–OH groups to increased UiO–66 (Zr) stability [65]. Hui [66] discovered that after heating to 300°C, the surface area of UiO–66 (Zr) was 1180–1240 m^2/g. This was attributed to the methylation benzene molecule size as it reaches the UiO–66 (Zr) structure. UiO–66 (Zr) toughness was evaluated by physical qualities such as breakdown above 500°C, stability after washing in boiling water and heating in air to 300°C for 6 hours, resistance for many chemicals, and crystallinity retention after exposure to 10 tons/cm^2 external pressure [66].

FIGURE 14.1   Molecular structure of UiO–66(Zr).

The zirconium terephthalate–based MOF UiO–66 (Zr) has homogeneous and controllable pore diameters, larger surface areas, and active Zr–O clusters. Due to the strong Zr–O bond with a high coordination number of Zr (IV) in a wide range of solvents, it also has incredible hydrothermal stability with high resistance to 350°C. These characteristics showed that UiO–66 (Zr) is a competitive future adsorbent [67]. Apart from that, UiO–66 (Zr) has high values and comprehensive characteristics that have led the investigators to accomplish the study using MOF. UiO–66 (Zr) is a suitable representative as porous materials to analyse the applicable properties and modification of UiOs [67]. The pore volumes for the UiO–66(Zr) was 0.91 m^3/g. UiO–66(Zr) MOF exhibits higher BET surface area and pore volumes [22].

The equilibrium time of UiO–66(Zr) was reported as 8 minutes for the adsorption of the BPA. The quantity of BPA adsorbed expanded with time until it reached a specific amount where no more BPA could be recovered from the solution. This phenomenon occurred due to number of surface sites accessible for adsorption during the early phases. After several times have elapsed, the remaining surface spots were difficult to occupy due to repulsion between the solute molecules of the solid and bulk phases. It was clear from Figure 14.2 that for 90 mg/L BPA solution, rapid adsorption occurred at the beginning of the process until equilibrium was established for UiO–66(Zr). In addition, UiO–66(Zr) and BPA complexes have possible van der Waals force, hydrogen, and smaller electrostatic interactions of the adsorbing pollutants. Thus, the highest uptake of the BPA recorded with 99.25% by the UiO–66(Zr) was due to the availability of active adsorption sites on the surface of the MOF and probably occurred primarily in the presence of functional groups of the UiO–66(Zr). In addition, the adsorption efficiency of UiO–66(Zr) onto removal of BPA also demonstrated great regeneration and reusability features with 74.9% in the

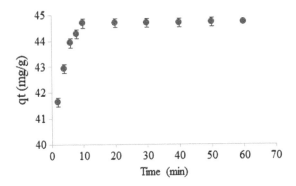

FIGURE 14.2 The equilibrium time for adsorption of BPA using UiO–66(Zr) MOF.

fourth cycle. This shows that UiO–66(Zr) is environmentally friendly and could save maintenances costs.

14.8 CONCLUSION

Among various methods available for BPA removal, adsorption using MOF was useful, simple, and significant. The possibility of employing UiO–66(Zr) for the green removal of BPA was discussed in this review. It was a great choice to improve the performance of the BPA removal by adsorption with the MOF adsorbents. The properties of the MOF can enhance the adsorption efficiency performance. However, MOF still needs to improve: the instability of the organic ligands in the MOF at high temperature and the validation of experimental design of the adsorption studies may be performed by an artificial neural network (ANN), response surface methodology (RSM), and the fittings of the models for statistical analysis. Thus, selecting an appropriate method to study the adsorption using MOF is important. This review suggested the potential adsorbent of UiO–66(Zr) as conventional adsorbents for the green adsorption of BPA from the wastewater because of its good performance of removal efficiency rate. The use of adsorption was interesting due to its rapid adsorption removal, reusable and regenerated, low cost, and practicality to the industrial scale.

ACKNOWLEDGEMENT

This work was sponsored by a grant provided through the financial aid from Universiti Teknologi PETRONAS (UTP) and the Ministry of Higher Education Malaysia, FRGS No: FRGS/1/2011/SG/UTP/02/13. The authors acknowledge the use of the facilities within the UTP Centralised Analytical Laboratory.

REFERENCES

[1] I. F. de Souza and D. F. Petri, "β–Cyclodextrin hydroxypropyl methylcellulose hydrogels for bisphenol A adsorption," *Journal of Molecular Liquids*, vol. 266, pp. 640–648, 2018.

[2] T. Encarnação, A. A. Pais, M. G. Campos, and H. D. Burrows, "Endocrine disrupting chemicals: Impact on human health, wildlife, and the environment," *Science Progress*, vol. 102, pp. 3–42, 2019.

[3] M. A. Ahsan, V. Jabbari, M. T. Islam, R. S. Turley, N. Dominguez, H. Kim, *et al.*, "Sustainable synthesis and remarkable adsorption capacity of MOF/graphene oxide and MOF/CNT based hybrid nanocomposites for the removal of Bisphenol A from water," *Science of the Total Environment*, vol. 673, pp. 306–317, 2019.

[4] A. Feigin, I. Ravina, and J. Shalhevet, *Irrigation with Treated Sewage Effluent: Management for Environmental Protection*, vol. 17. Springer Science & Business Media, 2012.

[5] O. Achi, "Practical case study industrial effluents and their impact on water quality of receiving rivers in Nigeria," *Journal of Applied Technology in Environmental Sanitation*, vol. 1, pp. 75–86, 2011.

[6] T. Ramachandra and M. Solanki, "Ecological assessment of lentic water bodies of Bangalore," *The Ministry of Science and Technology*, vol. 25, pp. 96, 2007.

[7] N. Singh, "Exposure to bisphenol–A through excess use of polymer, with environmental toxicity," *Int. J. Sci. Res. Sci. Eng. Technol*, vol. 2 (3), pp. 454–457, 2016.

[8] A. Careghini, A. F. Mastorgio, S. Saponaro, and E. Sezenna, "Bisphenol A, nonylphenols, benzophenones, and benzotriazoles in soils, groundwater, surface water, sediments, and food: A review," *Environmental Science and Pollution Research*, vol. 22, pp. 5711–5741, 2015.

[9] A. Usman and M. Ahmad, "From BPA to its analogues: Is it a safe journey?," *Chemosphere*, vol. 158, pp. 131–142, 2016.

[10] A. Alsbaiee, B. J. Smith, L. Xiao, Y. Ling, D. E. Helbling, and W. R. Dichtel, "Rapid removal of organic micropollutants from water by a porous β–cyclodextrin polymer," *Nature*, vol. 529, p. 190, 2016.

[11] B. Sandoval, "Perspectives on FDA's regulation of nanotechnology: Eemerging challenges and potential solutions," *Comprehensive Reviews in Food Science and Food Safety*, vol. 8, pp. 375–393, 2009.

[12] P. Fenichel, N. Chevalier, and F. Brucker–Davis, "Bisphenol A: An endocrine and meta-bolic disruptor," *Annales d'endocrinologie*, vol. 74 (3), pp. 211–220, 2013.

[13] S. Chu, G. D. Haffner, and R. J. Letcher, "Simultaneous determination of tetrabromobisphenol A, tetrachlorobisphenol A, bisphenol A and other halogenated analogues in sediment and sludge by high performance liquid chromatography–electrospray tandem mass spectrom-etry," *Journal of Chromatography A*, vol. 1097, pp. 25–32, 2005.

[14] M. Xiaoying, Z. Guangming, Z. Chang, W. Zisong, Y. Jian, L. Jianbing, *et al.*, "Characteristics of BPA removal from water by PACl–Al13 in coagulation process," *Journal of Colloid and Interface Science*, vol. 337, pp. 408–413, 2009.

[15] H.–S. Chang, K.–H. Choo, B. Lee, and S.–J. Choi, "The methods of identification, ana-lysis, and removal of endocrine disrupting compounds (EDCs) in water," *Journal of Hazardous Materials*, vol. 172, pp. 1–12, 2009.

[16] J. Corrales, L. A. Kristofco, W. B. Steele, B. S. Yates, C. S. Breed, E. S. Williams, *et al.*, "Global assessment of bisphenol A in the environment: Review and analysis of its occurrence and bioaccumulation," *Dose–Response*, vol. 13, p. 1559325815598308, 2015.

[17] M. Zhou, Y.–n. Wu, J. Qiao, J. Zhang, A. McDonald, G. Li, *et al.*, "The removal of bis-phenol A from aqueous solutions by MIL–53 (Al) and mesostructured MIL–53 (Al)," *Journal of Colloid and Interface Science*, vol. 405, pp. 157–163, 2013.

[18] S. Xu, Y. Lv, X. Zeng, and D. Cao, "ZIF–derived nitrogen–doped porous carbons as highly efficient adsorbents for removal of organic compounds from wastewater," *Chemical Engineering Journal*, vol. 323, pp. 502–511, 2017.

[19] X.–Y. Liu, M. Huang, H.–L. Ma, Z.–Q. Zhang, J.–M. Gao, Y.–L. Zhu, *et al.*, "Preparation of a carbon–based solid acid catalyst by sulfonating activated carbon in a chemical reduction process," *Molecules*, vol. 15, pp. 7188–7196, 2010.

[20] D. Zhao, D. J. Timmons, D. Yuan, and H.–C. Zhou, "Tuning the topology and functionality of metal–organic frameworks by ligand design," *Accounts of Chemical Research*, vol. 44, pp. 123–133, 2011.

[21] P. E. Rosenfeld and L. Feng, *Risks of Hazardous Wastes*. William Andrew, 2011.

[22] Y. Huang, C. Wong, J. Zheng, H. Bouwman, R. Barra, B. Wahlström, *et al.*, "Bisphenol A (BPA) in China: A review of sources, environmental levels, and potential human health impacts," *Environment International*, vol. 42, pp. 91–99, 2012.

[23] F. M. Ghazali and W. L. W. Johari, "The occurrence and analysis of bisphenol A (BPA) in environmental samples 'a review'," *Journal of Biochemistry, Microbiology and Biotechnology*, vol. 3, pp. 30–38, 2015.

[24] J. Xu, W. Chen, L. Wu, R. Green, and A. C. Chang, "Leachability of some emerging contaminants in reclaimed municipal wastewater-irrigated turf grass fields," *Environmental Toxicology and Chemistry: An International Journal*, vol. 28, pp. 1842–1850, 2009.

[25] G. Mileva, S. L. Baker, A. Konkle, and C. Bielajew, "Bisphenol–A: epigenetic reprogramming and effects on reproduction and behavior," *International Journal of Environmental Research and Public Health*, vol. 11, pp. 7537–7561, 2014.

[26] A. Razak, Z. Ujang, and H. Ozaki, "Removal of endocrine disrupting chemicals (EDCs) using low pressure reverse osmosis membrane (LPROM)," *Water Science and Technology*, vol. 56, pp. 161–168, 2007.

[27] T. Yamamoto, A. Yasuhara, H. Shiraishi, and O. Nakasugi, "Bisphenol A in hazardous waste landfill leachates," *Chemosphere*, vol. 42, pp. 415–418, 2001.

[28] T. Suzuki, Y. Nakagawa, I. Takano, K. Yaguchi, and K. Yasuda, "Environmental fate of bisphenol A and its biological metabolites in river water and their xeno–estrogenic activity," *Environmental Science & Technology*, vol. 38, pp. 2389–2396, 2004.

[29] J. Rocha, L. D. Carlos, F. A. A. Paz, and D. Ananias, "Luminescent multifunctional lanthanides–based metal–organic frameworks," *Chemical Society Reviews*, vol. 40, pp. 926–940, 2011.

[30] T. M. Brooks, R. A. Mittermeier, G. A. da Fonseca, J. Gerlach, M. Hoffmann, J. F. Lamoreux, *et al.*, "Global biodiversity conservation priorities," *Science*, vol. 313, pp. 58–61, 2006.

[31] J. E. Cooper, E. L. Kendig, and S. M. Belcher, "Assessment of bisphenol A released from reusable plastic, aluminium and stainless steel water bottles," *Chemosphere*, vol. 85, pp. 943–947, 2011.

[32] K.–L. Chang, J.–F. Hsieh, B.–M. Ou, M.–H. Chang, W.–Y. Hseih, J.–H. Lin, *et al.*, "Adsorption studies on the removal of an endocrine–disrupting compound (Bisphenol A) using activated carbon from rice straw agricultural waste," *Separation Science and Technology*, vol. 47, pp. 1514–1521, 2012.

[33] S. Yüksel, N. Kabay, and M. Yüksel, "Removal of bisphenol A (BPA) from water by various nanofiltration (NF) and reverse osmosis (RO) membranes," *Journal of Hazardous Materials*, vol. 263, pp. 307–310, 2013.

[34] C. G. Moreira, M. H. Moreira, V. M. Silva, H. G. Santos, D. M. Bila, and F. V. Fonseca, "Treatment of Bisphenol A (BPA) in water using UV/H_2O_2 and reverse osmosis (RO) membranes: assessment of estrogenic activity and membrane adsorption," *Water Science and Technology*, vol. 80, pp. 2169–2178, 2019.

[35] J. Heo, J. R. Flora, N. Her, Y.–G. Park, J. Cho, A. Son, *et al.*, "Removal of bisphenol A and 17β–estradiol in single walled carbon nanotubes–ultrafiltration (SWNTs–UF) membrane systems," *Separation and Purification Technology*, vol. 90, pp. 39–52, 2012.

[36] L. Mita, L. Grumiro, S. Rossi, C. Bianco, R. Defez, P. Gallo, *et al.*, "Bisphenol A removal by a *Pseudomonas aeruginosa* immobilized on granular activated carbon and operating in a fluidized bed reactor," *Journal of Hazardous Materials*, vol. 291, pp. 129–135, 2015.

[37] Y. Onundi, B. A. Drake, R. T. Malecky, M. A. DeNardo, M. R. Mills, S. Kundu, *et al.*, "A multidisciplinary investigation of the technical and environmental performances of TAML/peroxide elimination of Bisphenol A compounds from water," *Green Chemistry*, vol. 19, pp. 4234–4262, 2017.

[38] Y. Xi, Z. Sun, T. Hreid, G. A. Ayoko, and R. L. Frost, "Bisphenol A degradation enhanced by air bubbles via advanced oxidation using in situ generated ferrous ions from nano zero–valent iron/palygorskite composite materials," *Chemical Engineering Journal*, vol. 247, pp. 66–74, 2014.

[39] I. Bautista–Toledo, M. Ferro–García, J. Rivera–Utrilla, C. Moreno–Castilla, and F. Vegas Fernández, "Bisphenol A removal from water by activated carbon. Effects of carbon characteristics and solution chemistry," *Environmental Science & Technology*, vol. 39, pp. 6246–6250, 2005.

[40] S. Zheng, Z. Sun, Y. Park, G. A. Ayoko, and R. L. Frost, "Removal of bisphenol A from wastewater by Ca–montmorillonite modified with selected surfactants," *Chemical Engineering Journal*, vol. 234, pp. 416–422, 2013.

[41] J. Kwon and B. Lee, "Bisphenol A adsorption using reduced graphene oxide prepared by physical and chemical reduction methods," *Chemical Engineering Research and Design*, vol. 104, pp. 519–529, 2015.

[42] Q. Yang, M. Gao, Z. Luo, and S. Yang, "Enhanced removal of bisphenol A from aqueous solution by organo–montmorillonites modified with novel Gemini pyridinium surfactants containing long alkyl chain," *Chemical Engineering Journal*, vol. 285, pp. 27–38, 2016.

[43] J. R. Koduru, L. P. Lingamdinne, J. Singh, and K.–H. Choo, "Effective removal of bis-phenol A (BPA) from water using a goethite/activated carbon composite," *Process Safety and Environmental Protection*, vol. 103, pp. 87–96, 2016.

[44] V. R. Moreira, Y. A. Lebron, R. F. Gomes, B. F. Tatiane de Paula, B. G. Reis, L. V. Santos, *et al.*, "Enhancing biodegradability and reducing toxicity of a refinery wastewater through UV/H_2O_2 pretreatment," *Journal of Environmental Chemical Engineering*, vol. 8, p. 104442, 2020.

[45] K. K. Choy, G. McKay, and J. F. Porter, "Sorption of acid dyes from effluents using activated carbon," *Resources, Conservation and Recycling*, vol. 27, pp. 57–71, 1999.

[46] M. Özacar and İ. A. Şengil, "Adsorption of reactive dyes on calcined alunite from aqueous solutions," *Journal of Hazardous Materials*, vol. 98, pp. 211–224, 2003.

[47] C.–C. Lee, L.–Y. Jiang, Y.–L. Kuo, C.–Y. Hsieh, C. S. Chen, and C.–J. Tien, "The poten-tial role of water quality parameters on occurrence of nonylphenol and bisphenol A and identification of their discharge sources in the river ecosystems," *Chemosphere*, vol. 91, pp. 904–911, 2013.

[48] K.–L. Chang, J.–F. Hsieh, B.–M. Ou, M.–H. Chang, W.–Y. Hseih, J.–H. Lin, *et al.*, "Adsorption studies on the removal of an endocrine–disrupting compound (Bisphenol A) using activated carbon from rice straw agricultural waste," *Separation Science and Technology*, vol. 47, pp. 1514–1521, 2012.

[49] P. Benard and R. Chahine, "Determination of the adsorption isotherms of hydrogen on activated carbons above the critical temperature of the adsorbate over wide temperature and pressure ranges," *Langmuir*, vol. 17, pp. 1950–1955, 2001.

[50] G. Crini, E. Lichtfouse, L. D. Wilson, and N. Morin–Crini, "Conventional and non–conventional adsorbents for wastewater treatment," *Environmental Chemistry Letters*, vol. 17, pp. 195–213, 2019.

[51] K. Noufel, N. Djebri, N. Boukhalfa, M. Boutahala, and A. Dakhouche, "Removal of bisphenol A and trichlorophenol from aqueous solutions by adsorption with organically modified bentonite, activated carbon composites: A comparative study in single and binary systems," *Groundwater for Sustainable Development*, vol. 11, p. 100477, 2020.

[52] Y. Zhou, J. He, J. Lu, Y. Liu, and Y. Zhou, "Enhanced removal of bisphenol A by cyclodextrin in photocatalytic systems: degradation intermediates and toxicity evaluation," *Chinese Chemical Letters*, vol. 31, pp. 2623–2626, 2020.

[53] D. Mohanta and M. Ahmaruzzaman, "Biogenic synthesis of SnO_2 quantum dots encapsulated carbon nanoflakes: An efficient integrated photocatalytic adsorbent for the removal of bisphenol A from aqueous solution," *Journal of Alloys and Compounds*, vol. 828, p. 154093, 2020.

[54] J. Liu, H. Qiu, F. Zhang, and Y. Li, "Zeolitic imidazolate framework–8 coated Fe_3O_4@ Si_2 composites for magnetic solid–phase extraction of bisphenols," *New Journal of Chemistry*, vol. 44, pp. 5324–5332, 2020.

[55] C.–S. Cao, J. Wang, X. Yu, Y. Zhang, and L. Zhu, "Photodegradation of seven bisphenol analogues by Bi_5O_7I/UiO–67 heterojunction: Relationship between the chemical structures and removal efficiency," *Applied Catalysis B: Environmental*, vol. 277, p. 119222, 2020.

[56] Y. Liu, Y. Wang, Q. Wang, B. Wang, X. Liu, and B. Wu, "Adsorption and removal of bisphenol A in two types of sediments and its relationships with bacterial community," *International Biodeterioration & Biodegradation*, vol. 153, p. 105021, 2020.

[57] Z. Katančić, W.–T. Chen, G. I. Waterhouse, H. Kušić, A. L. Božić, Z. Hrnjak–Murgić, et al., "Solar–active photocatalysts based on TiO_2 and conductive polymer PEDOT for the removal of bisphenol A," *Journal of Photochemistry and Photobiology A: Chemistry*, vol. 396, p. 112546, 2020.

[58] S. Sadeghzadeh, Z. G. Nejad, S. Ghasemi, M. Khafaji, and S. M. Borghei, "Removal of bisphenol A in aqueous solution using magnetic cross–linked laccase aggregates from Trametes hirsuta," *Bioresource Technology*, vol. 306, p. 123169, 2020.

[59] O. M. Yaghi, M. O'Keeffe, N. W. Ockwig, H. K. Chae, M. Eddaoudi, and J. Kim, "Reticular synthesis and the design of new materials," *Nature*, vol. 423, pp. 705–714, 2003.

[60] W. Lu, Z. Wei, Z.–Y. Gu, T.–F. Liu, J. Park, J. Park, et al., "Tuning the structure and function of metal–organic frameworks via linker design," *Chemical Society Reviews*, vol. 43, pp. 5561–5593, 2014.

[61] S. Lowell and J. E. Shields, *Powder Surface Area and Porosity*, vol. 2. Springer Science & Business Media, 2013.

[62] F. S. Baker, C. E. Miller, A. J. Repik, and E. D. Tolles, "Activated carbon," *Kirk-Othmer Encyclopedia of Chemical Technology*, John Wiley & Sons, New York, NY, USA, vol. 4, pp. 1015–1037, 1992.

[63] M. Ahmedna, W. E. Marshall, A. A. Husseiny, R. M. Rao, and I. Goktepe, "The use of nutshell carbons in drinking water filters for removal of trace metals," *Water Research*, vol. 38, pp. 1062–1068, 2004.

[64] S. E. Bambalaza, H. W. Langmi, R. Mokaya, N. M. Musyoka, J. Ren, and L. E. Khotseng, "Compaction of a zirconium metal–organic framework (UiO–66) for high density hydrogen storage applications," *Journal of Materials Chemistry A*, vol. 6, pp. 23569–23577, 2018.

[65] L. Valenzano, B. Civalleri, S. Chavan, S. Bordiga, M. H. Nilsen, S. Jakobsen, et al., "Disclosing the complex structure of UiO–66 metal organic framework: A synergic combination of experiment and theory," *Chemistry of Materials*, vol. 23, pp. 1700–1718, 2011.

[66] H. Wu, T. Yildirim, and W. Zhou, "Exceptional mechanical stability of highly porous zirconium metal–organic framework UiO–66 and its important implications," *The Journal of Physical Chemistry Letters*, vol. 4, pp. 925–930, 2013.

[67] D. Zou and D. Liu, "Understanding the modifications and applications of highly stable porous frameworks via UiO–66," *Materials Today Chemistry*, vol. 12, pp. 139–165, 2019.

Index